VOICES
FROM THE
SHORELINE

THE ANCIENT &
INGENIOUS TRADITIONS
OF COASTAL FISHING

T0323353

MIKE SMYLIE

FOREWORD BY MARK HORTON

The History Press

*To all those Voices, past and present,
and Ana & Otis for sharing much of
the journey.*

Salmon

First published 2021

The History Press
97 St George's Place, Cheltenham,
Gloucestershire, GL50 3QB
www.thehistorypress.co.uk

© Mike Smylie 2021

The right of Mike Smylie to be identified as the Author
of this work has been asserted in accordance with the
Copyright, Designs and Patents Act 1988.

British Library Cataloguing in Publication Data.
A catalogue record for this book is available from the British Library.

ISBN 978 0 7509 9755 3

Typesetting and origination by The History Press
Printed and bound in Great Britain by TJ Books Limited, Padstow, Cornwall.

Fish Pattern © Freepik.com

Trees for Life

CONTENTS

FOREWORD

The western coasts of the British Isles are among the most spectacular and diverse in the world – from the cliffs of Cornwall, the mudflats of the Severn Estuary, the rocky Welsh coasts and the spectacular scenery of Solway Firth. But these are not uninhabited spaces, but places, where for generations fishermen have sought out their living, following the herring and netting the salmon. Each place generated its own unique way to harvest the sea and hand down this knowledge, most probably over millennia.

Sadly, many of these traditional fisheries now lie abandoned, as fish stocks have dwindled and zealous government rules are being arbitrarily imposed so the fishermen are no longer able to continue. We are most likely to be the last generation to be able to observe one of Britain's most ancient and intangible heritages.

This book will help to make a record of this disappearing heritage. Mike Smylie has spent a lifetime recording these fishing communities, photographing their craft and interviewing the practitioners. Sadly, many of the fishermen have now retired or died, and their know-how with them; others just hang on, trying to pass on the traditions to the next generation. But we all fear that the golden thread of knowledge will be lost forever and only this written record will survive.

I write this foreword in what was one of the main salmon fishing ports of the Severn, where thousands of fish were landed every year and have been since the Domesday Book. I am looking out at the wrecks of three stop-net boats, covered in brambles, which last ventured to sea in the 1990s. In ten years, even these will be gone, along with the salmon, and with them most likely the remaining traditional fisheries of western Britain. I hope that this book will be a call to arms, for everyone who is passionate about our coasts, that this precious heritage can be preserved and passed on to future generations.

Mark Horton
Maritime and historical archaeologist
Gatcombe, Gloucestershire

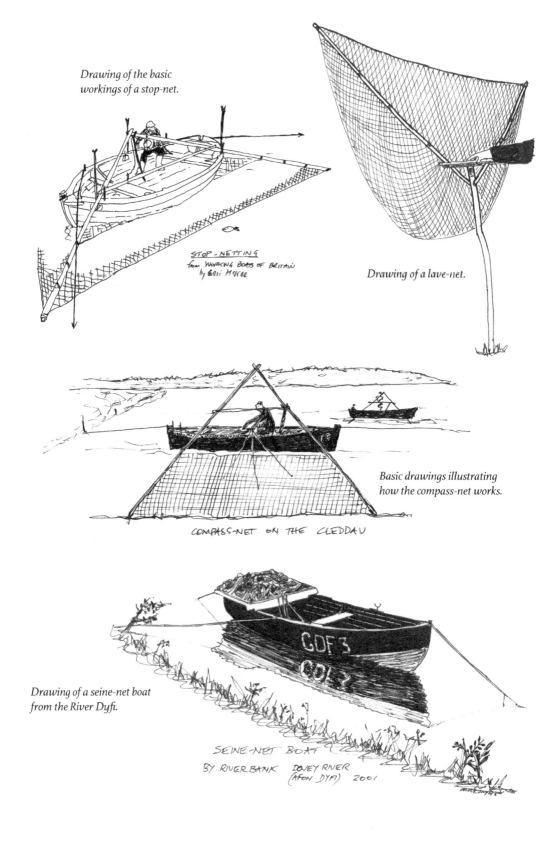

Drawing of the basic workings of a stop-net.

STOP - NETTING
from WORKING BOATS OF BRITAIN
by Eric McKee

Drawing of a lave-net.

Basic drawings illustrating how the compass-net works.

COMPASS-NET ON THE CLEDDAU

Drawing of a seine-net boat from the River Dyfi.

GDF 3

SEINE-NET BOAT
BY RIVERBANK DOVEY RIVER
(AFON DYFI) 2001

JOURNEY BEGINNING

Salmon and herring. Two great fish that have their own histories wrapped up in and around the evolution of what we know as Britain. One the King of Fish, the other the King of the Sea; one the Silver King, the other the Silver Darling. Then there's the unprincely shrimp, the poor relative, neither regal nor silvery, and why will become clear!

I set out to record the shoreline fisheries of the west coast of Britain, my coast, so to speak. One I think I know best, the one I have lived nearby most of my life. Sometimes right by it, at other times upon it while residing aboard various boats. The coast is the fringe between land and sea, a boundary of importance. Some call it liminal, a border or bridge between life and death, a place to worship. In the Iron Age treasure was buried here, sometimes even bodies. To others it is simply a threshold where solid ground gives way to immersion.

Nevertheless, this is a coast that over a good part is pretty wild and untamed, edged against some seas that have not, in the past, been thought of as the richest in sea life: the Bristol Channel, the Irish Sea, the two firths of the Solway and the Clyde, those indented Hebridean lochs and the wild Minch. Others will say, on the contrary, that the west coast is diverse. One fellow told me he thought the Irish Sea has the largest mixed-species fishery in the country, while the amount of herring boats working the west coast up to the 1950s suggests a huge fishery. Today, prawns, shrimps, scallops and mussels account for the most landings but things have changed in the nature of the ways of fishing. Salmon and herring have almost disappeared from this coast due to various pressures: climate change, overfishing and pollution.

This is the coast that I care most about, and equally I chose to support those that have, over generations, made a living by it in sustainable ways. In most cases, these folk have sought two fish as they were the most profitable: salmon and herring. Consequently, I chose to confine my research, to limit it to just these two great fish. I didn't really want to go into deep water (literally and metaphorically speaking) because that has been done numerous times. What I wanted was to take the reader on a discovery of the methods that have been used throughout generations of coastal fishing, and at the same time to explore the communities that have learned to sustain their fishing through consistency between those generations. Where people have grown up in a community and stayed there, the elders teaching the next generation the value of tradition. But then technology arrived to destroy much. As one old fellow told me, 'They should never have allowed the boats to get so big.'

There are still a few remaining open avenues to earn a living by fishing that do not necessitate the huge investment in fishing vessels and the associated gear. Small is beautiful, they say, and to stand facing the ocean holding a net, lave- or haaf-, is simply as basic as shrimping in rock pools the way children do it today, and have done so for centuries.

I pause. Much is said about the romanticising of fishing, of fishermen living the ideal life, earning a crust from their chosen profession, sailing out and hunting the shoals. Whereas that 'profession' might be used to describe many fishermen of today, there is little romantic notion in the job itself. It's tedious and monotonous, hard work and often offering little in reward for that labour, dirty and smelly when covered in fish guts and scales, and dangerous when relying upon the vagaries of British weather. Having said that, many folk I've spoken to talk about fishing in a way that forms a beautiful picture in my brain. They are respectful of the job, and respectful of the memory of those that didn't survive to the end of their working days.

But hang on a minute. 'You are talking about fishermen who go off to sea for a week or two at a time,' you say, 'whereas here you are with fishermen who simply work the tides from their home.' And mostly you are quite correct in your surmising. But to work the tides also means working hours that do not obey the rules of daylight. Maybe spending a week on a trawler out in the Western Approaches (which I've done) is somewhat different to time spent at home afterwards, but fishing the shoreline means fishing every tide when possible and if that means standing out in a river at three o'clock in the morning on your own, with a Force 8 gale blowing, then that is what they all do.

My initial intention was to undertake a journey from Appledore in northern Devon, all the way up this coast to Applecross, some 100 miles short of Cape Wrath. Why choose these two villages? Ignoring the name association,

I thought that the coast in between would give me a good insight to what was happening fish-wise. North Cornwall was never a base for commercial salmon fishing, excepting the odd fish that swam into the rivers such as the Camel, but nothing of substance. It is not much of a herring fishing coast either, although it is worth mentioning that the rivers Tamar, Lynher and Dart, for example, all flowing into the English Channel, were home to some commercial salmon fishing. But the estuary of the rivers Taw and Torridge at Appledore were noted salmon-netting areas.

Then Applecross, a hidden gem over the infamous 2,053ft-high Bealach na Bà road, was once noted for its salmon fishing, even though most of the signs of it are now gone. Applecross – meaning 'sanctuary' in its Gaelic form – is actually a conglomeration of various townships in the area, and, although it was eventually somewhat disappointing in terms of fishing traditions, we did search along the coast to the south – around Milltown, Camustiel, Camusterrach, Culduie and Toscaig. The nearest anyone knew about in the locality were the bag-nets over at Raasay, and they were long gone. Although visiting is most pleasurable, we had to venture further north to find more commercial salmon history!

You see that was the initial idea from the outset: to study salmon and the various fishing methods undertaken both in the past and still today (just) for this prized fish. It was not to be a potted history of salmon as many others have written about this fish, its life cycle and history. I did, in no way, wish to repeat any of these. No, this was to be practical as much as possible, a chance to wet my toes and catch a glimpse of traditional ways of the past.

The exciting thing about salmon fishing, as against other fish, is the multitude of differing manners in which they are captured. Evocative names such as the lave-net, compass-net, whammel-net, sweep-net, haaf-net, bag-net, seine-net and fly-net conjure up so many images of what is actually a labour-intensive occupation, often with little remuneration. But there are so many more terms given to the various ways that have been utilised over time to capture salmon: the cruive, the yair or fishing stank, the mud-hangs, the stage-net, the toot-, cairn-, croy-, pot-, draw-net, hang-, bob-, T-net, teedle-, stake-, kettle-, keddle-, stell-, drift-net, poke-, paidle-net, paddle-, sparling-, cleek-, long-, tide-, floating-, jumper-net, even wear-shot-net, and all are variations on several types while, at the same time, being methods for catching salmon. Not necessarily just for salmon, but certainly effective at fishing for it. None of your boring old trawling here!

'Herring' from Jonathan Couch's A History of the Fishes of the British Islands, *1865.*

But then herring dropped into the frame and, by comparison, catching this fish could be said to be pretty boring too. Drift-netting and ring-netting were the only ways by which the shoals were captured until the advent of the modern trawling, pair-trawling and purse-seining, all of which devastated the stocks; and we ain't going there! Drift-netting was practised all around Britain and further afield, and is the age-old method. Ring-netting, confined on the whole to the west coast (OK, yes, there was some on the east side but nothing like the west), was something new in the 1850s, and over the next century gained much in stature in some places while it was vilified in others. Today, although there are hardly any herring to ring and the method dwindled in the 1970s to an eventual stoppage, there remains much folklore and nostalgia among many fishermen. I didn't want to encroach into Angus Martin's fine *The Ring-Net Fishermen*, so I asked him for advice and found another course to navigate.

There are, of course, other fish, considered by some to be of a similar stature to salmon and herring, but I leave those to writers of superior literary skill, with their high-flying contacts and double-barrelled surnames, who seem to dip intermittently into fishery history when they appear to need a subject to write about. Their writing is, as expected, swift and sweet, but lacks any passion for commercial fishermen. Passion, I'm told, is what makes the subject alive, and coastal fishermen certainly do have my, and many others', hearts. Passion creates possibility.

Shrimps jumped on the bandwagon at a later stage, partly through an invitation to sample a spot of net clearing in Weston-super-Mare, and partly because of my old friend Tom Smith of Sunderland Point. He fished for shrimps with his horse over the sands of Morecambe Bay, in the same way as folk had done for generations. How could I not include such a meaningful and redolent imagery of the sands?

And so, in this journey, we start in Appledore, and eventually arrive in Applecross, but, as I say, we do need to travel slightly further. We will just have to see what occurs. Just as the old fisherman once said, 'Sometimes I sit and think … sometimes I just sit', I often think that sometimes I drive and think (with an end in mind) … and sometimes I just drive!

Maybe this is the time to say something about these three wonderful species that, alongside those whose voices we hear, are the heroes of this book. Herring swim in vast shoals and deposit their sticky eggs on coarse sand, gravel, shells and small stones, all the members of a shoal spawning over a relatively short time period. Around our coasts the fish tend to congregate on traditional spawning grounds, many of which are on shoals and banks and in relatively shallow water, approximately 15–40m deep. The Ballantrae Banks in the southern Firth of Clyde are a good example. Once the small fish develop, they tend to stay in shallow water for a couple of years before swimming out to join the adult population in the north-east Atlantic – the vast area between Scotland, Iceland and Norway. Then to spawn they swim southwards, flooding into the southern parts of the North Sea and down the west side of Britain and Ireland. Some are anadromous, in that they spawn in rivers and then return to the sea, while others return to the grounds where they were bred. They tend to stay in deep water throughout the day and come to the surface at night to feed off the plankton.

Salmon have similar habits, although they swim further. It is an incredible fish that, once matured, swims down river and out into the north Atlantic, perhaps as far as Greenland, to return to spawn in the exact same bit of clean gravel, in the exact same steadily flowing stream where they had hatched out of eggs laid in their mother's redd. As 'alevins' they hung about, feeding off their yolk sac, growing into 'fry', then foraging for food in that stream. Over a period of up to five years they remained in the stream, becoming then 'parr', then later 'smolts' as they prepared to swim downstream and out into the ocean. Anadromous they are too, living first in fresh water, then acclimatising to saltwater.

Salmon have, over time, had more written about them than any other fish, perhaps due to their position as the 'king of all fish'. Furthermore, no fish has been legislated about as much as he, and probably no fish has created the same profitability among the fishermen of northern Europe, though surely the herring must come a close second, probably followed by cod.

The trade in both salmon and herring is indeed ancient. Various records on herring date back to the sixth and seventh centuries, and salmon almost as old, though it is also known from evidence from remains that both fish were favoured by the Romans. Let's face it, the taking of salmon and herring from the sea probably dates back to when man first started fishing, or indeed finding fish stranded in rock pools, when the building of fish weirs followed.

Both these fish come in various sizes, depending on all manner of factors such as being members of different sub-species, their age and where they were caught. Big Atlantic herring are huge compared with the small Clovelly herring. We shall hear of grilse too, an Atlantic salmon that has spent only one winter at sea before returning to the river. Salmon grilse are often difficult to distinguish from salmon that have been at sea longer, except by scale reading. They tend to be smaller on average (2–3lb in May, 5–7lb in July) but when they enter our rivers in September, often attain 8–10lb and in October 12–15lb. Or so I'm told.

The very same rock pools would have been full of shrimps, though things have moved on since the Romans and their forefathers were poking around them. Generally, it is the brown shrimp that lives close inshore and the pink shrimp that is found in deeper water, and that is how it is still fished in some quarters. Fishing with a horse restricts the fisher to the shallow water and thus it is the brown shrimps that today's tractors are after, although others go out onto the sands with their push nets 'putting' for shrimps. It's the boats that go for the pink ones.

The desire for shrimp teas in Victorian Britain grew as nineteenth-century factory workers' rights included sensible holidays so that they spread out from the industrial mills and factories and decamped upon the up-and-coming coastal resorts at weekends and these vacation times. Potted shrimps – shrimps in butter – also soon became a delicacy where shrimps were fished and remains so today, especially in the north-west.

But today we all know that fish stocks are under pressure, and nowhere is this more obvious than with the Atlantic salmon. Recent figures show this decline. In the southern area of the north-east Atlantic (comprising Ireland, the UK, France and the southern part of Iceland), grilse have declined by 66 per cent since 1970 and multi-sea winter fish ('springers') by 81 per cent. Measures have been put in place over the last few decades but no one is sure that they are working. For example, in 2007, the Irish authorities banned mixed stock drift-netting. However, in 2006, the last year of drift-netting, the reported catch in that fishery was 70,000 fish and, after making a modest adjustment for the extensive illegal catch in this fishery, probably at least of the order of 100,000 fish. Notwithstanding the 'saving' of this very large number of salmon, neither the commercial estuary nets nor the angling effort have since exceeded their pre-2007 catch levels.

This is shown through statistics: adult salmon returning to the Irish coast has collapsed from 2 million in the late 1970s to 250,000 in recent years (2017). Furthermore, the marine survival of the species had fallen from 20 per cent of fish returning to rivers in 1980 to 5 per cent in 2017, before any exploitation. Catches of salmon in Ireland were at an all-time low then, and still are, at about 22,000 fish by all methods (comprising 7,000 by commercial nets and 15,000 by angling). At the same time, it was said that angling had accounted for 63 per cent of salmon catches since 2007. This mirrors the case both sides of the Irish Sea, and, in respect of this book, majorly affects the west coast. The general consensus is that this is due to all manner of causes; climate change being the prime reason, although add on agricultural pollution off the land, exploitation, increased juvenile predation in freshwater, migration barriers, the increase in seal populations, open-cage salmon farming and habitat degradation. Overfishing would undoubtedly have been added up to recent times, but with the tiny number of commercial fishermen forced out, stocks still don't seem to be improving.

'Salmon' from John Couch's A History of the Fishes of the British Islands, *1865.*

One question I hear often is, 'Will wild salmon return?' As was pointed out to me by Douglas Carson, no one seems to have researched the effect on salmon stocks when Iceland and Greenland fishermen discovered a huge market for the fish. Did this coincide with the sea temperature increase due to climate change? Did the drift-netters working off the Irish coast up to the beginning of this century make inroads into stocks? The answer, from common sense, is probably yes but nothing is certain. What is certain is that the wild salmon stocks are low. Compared with Pacific salmon, where the fishery is well managed, especially in Alaska, then what lessons can we learn here?

Some regard the Pacific salmon as aquaculture because the fish are put into rivers from the hundreds of hatcheries up Canadian and west coast US rivers. Depending on which species (there are generally five: Chinook, Sockeye, Coho, Chum and Pink), some can be released after a few days and others after eighteen months of growth. Some 900,000 fish are landed a year and sold, and it is estimated that 34 per cent of these are from hatcheries. One mature adult fish is said to be caught for every ten released from a hatchery, so that nature takes its percentage. Yet that leaves 66 per cent from the wild, which surely suggests success. On the other hand, are we in the UK too wrapped up in supporting the farming of salmon in pods, with its knock-on effect on jobs, to be bothered? Meanwhile, herring continues to be plundered in quantity and fed to the farmed salmon because to us, the general populace, this fish has fallen out of favour. Surely there's some irony there!

On the beaches, the same generations of children continue their exciting researches into the summer's rock pools with their shrimping nets without really realising what they have. Tractors make life on the sands less labour intensive but perhaps more dangerous than they were. Nevertheless, salmon, herring and shrimps are still very much with us in some form, and will, if we take some care for this planet of ours, remain so for the foreseeable future.

It's worth mentioning that this book was beginning to consolidate into a narrative back in March 2020, at a time when the first coronavirus lockdown was thrust upon us. Since then, over the year, it has been almost impossible to interview elderly fishermen without the possibility of me bringing contamination upon them, always a risk even with the strictest precautions. Thus some interviews haven't been concluded and others have been done by phone. This has sometimes been difficult, even impossible in one case when the particular fisherman's hearing aid had broken and he couldn't hear me, even with me shouting down the handset. To make matters worse, he couldn't get out to obtain a replacement, which was so upsetting and confusing for him as he was in his early nineties. Yes, it's been confusing all round, with the virus, Brexit and a downturn in economic activity. Sometimes it's surprising that there's any fishing industry left in this country, given successive governments', and their quangos', total indifference to it.

Finally, I generally use the term 'fisher' for sport fishers, anglers, 'the rod brigade' (as some call them), while commercial fishermen are exactly that, or 'fisherfolk', sometimes 'fishing people'. Nothing untoward, simply a matter of distinction!

Brendan Sellick, whom we meet in chapter 4, sadly died at the age of 86 during the production stage of this book. Doubt lingers whether his son Adrian will continue to fish without his dad's support ...

PART 1

SOUTH-WEST
ENGLAND

1

APPLEDORE SALMON

You'd be excused for missing Appledore when you drive exhilarated down the A39, northern Cornwall beckoning in the distance. In the excitement you might simply shoot on southwards, while others might be tempted away on seeing a signpost to Kingsley Amis' Westward Ho! But Appledore has had its moments, nestled as it is at the confluence of the rivers Taw and Torridge, and is often described, as many others are, as a fishing village with narrow streets and colourful houses. Indeed, you do have to wonder what, without the description of 'a colourful village with olde worlde cobbled streets and oozing with character', many of our best-known coastal tourist traps would be!

However, in support of Appledore, it does have one wonderful atmospheric narrow street, edged on both sides by a variety of different-shaped fishermen's houses, although they weren't as brightly contrasting in colour as some of the photos I'd seen. Those on the seaward side came with the added bonus to fishers that they have direct access to the river frontage. The Beaver Inn, halfway along, has been serving the fishing community, as well as sailors from all over, for over 400 years and it is said that press gangs worked here, wearing beaver pelt hats that gave the pub its name and pressing the King's Shilling into the hands of unsuspecting drinkers. Another suggestion is that the ships coming into Appledore brought beaver pelts with them. Nevertheless, the salmon boats were also moored alongside the quay just below. Standing outside the pub, gazing seaward, it is easy to see just why Appledore is perfectly placed, geographically speaking.

With the wider River Taw heading north-east, then east, to Barnstaple, and the Torridge heading south, Appledore is right at the confluence, on its south side, facing Crow Point to the north. With a long quayside to receive the numerous ships from Wales, Ireland, France and Spain, it was also equally well placed for the salmon fishers to work the tidal waters from where the rivers join and empty into Barnstaple Bay, and upstream upon both rivers.

Looking east across the Torridge reveals an expanse of sands, dotted with boats, some sailing, and the jumbled line of houses below the green hills; this is Instow. With its railway from Barnstaple bringing in tourists until its closure in 1985, it became a haven of activity, though now, I admit, it retains a more peaceful air. From across the river, with the sun shining on the golden sands, it sure looks the part. But there was fishing here, too, as there was upstream of the Taw at Fremington Quay, which, once the railway had been built in the 1840s, became a busy port, largely due to the Taw silting up and boats not being able to reach Barnstaple. Exports were mostly ball clay from the local clay pits.

A maritime port needs boats and ships, so it's no surprise to find that Appledore was also home to several boatbuilders. Some of its ships sailed to Newfoundland to fish back in the sixteenth century, as did many West Country vessels, catching cod that was then dropped off in northern Spain before they returned home with wine and other badly needed goods, with the cycle recommencing the following spring.

Among the fine array of boatyards was J. Hinks and Sons, which had been building wooden fishing boats since the 1840s, and thus there was no shortage of craft for the local fishers, both for inshore and river fishing.

Although Alan Hinks, who took over the firm in the 1960s, continued building 70ft boats for Scottish fishing owners, by the 1990s orders had dried up and the yard had closed. Shipbuilding must have exercised some blood control, though, as Appledore Shipyard, again dating from 1855, only recently closed in March 2019, though they generally built larger boats than fishing vessels throughout their existence. The current thinking is that it will reopen to build vessels for the Faroese fishing industry. Another well-known name was that of P.B. Waters and Sons, a company stemming from Tom Waters, who had moved from Clovelly to East Bideford in around 1855 and then later to Appledore; but we aren't here to talk about boatbuilding, or indeed cod; it is salmon, that King of Fish, that we've started our search here for.

So we find that both the rivers, Taw and Torridge, were good salmon rivers, just as the estuary was, once both rivers had joined it, and the area has been fished for many centuries. I guess 'were' and 'was' are the operative words as some say there's plenty of salmon in the river. It's just that no one but the anglers are allowed to catch it.

Hauling in at the Bridge Pool, Bideford. (North Devon Maritime Museum)

They have, without doubt, been good salmon rivers, so far back that nobody really knows when the first salmon was caught. Common sense tells me it was probably when man first realised he could eat fish, long before the first mention of it in the Domesday Book! But as his fishing methods evolved, at some point we arrived at the era of the seine-net.

The seine-net is probably the oldest method of netting and is rowed out into the river upon a small open boat – or off a beach directly into the ocean as often as not, though never in these parts. One end is held to the shore while the boat travels in a circle, shooting the net as it goes, to arrive back at shore. On these two rivers a place that a seine is set out is called a 'draft', which often as not is little more than a pool of deeper water within the river at low water where a net can be laid out by boat. Once the net is out, the two ends are brought together and are then hauled in. Sounds easy but, given strong and erratic tides and currents, heavy nets, uncertain riverbeds and any other number of variables, it certainly is not.

On these rivers the net was worked just before low water, then through slack water and into the first of the flood. It took a couple of hours, sometimes more, depending on the tide. That meant the fishers could work two tides a day. Generally speaking, the Taw was fished by people from Barnstaple and Braunton, while the Torridge was the domain of fisherfolk from anywhere between Appledore and Bideford. Fishing above both downstream bridges across each river seldom occurred much in the early decades of the twentieth century.

Then there's the legislation that necessitated having a licence. Before the 1860s, fishermen set their nets unfettered by any regulation and one of the first hereabout was to ban Sunday fishing. This probably marked the beginning of years of conflict between the fishermen and the legislators, who most thought were on the side of the rod fishers. Today, those same sentiments linger on and we shall be hearing much more about them as we move around the various fishing communities. This is obvious in all the various small fishing communities that once relied on this sort of fishing. What we shall see over the coming pages is that the conservation of fish is not the main driving force behind the loss of commercial fishing (although it certainly plays a part and sometimes seems to be an excuse drummed up by the authorities), but finance. In the early days of legislation, those conservators, who were appointed by Justices of the Peace, were gentlemen who enjoyed sport fishing with rods, much like the JPs themselves, so it was pretty obvious whom the regulations were going to favour. These days, it is those very same anglers fishing with rods that pay far more through various licences into the coffers of the various government agencies (Environment Agency (EA), National Resources Wales (NRW), Marine Scotland etc.) than do most commercial fishermen, excluding those working in the aquaculture industry.

So, prior to the 1860s, any amount of fishing gear could be set in rivers and estuaries and vast amounts of salmon were caught. It was almost a case of there being a limitless amount and it was eaten abundantly in the growing towns and cities. The general consensus was that folk were tired of it, and word got around that the service sector was rebelling being fed on it: it is said that they insisted they must only be fed salmon three times a week. This tale does change somewhat and there's no documented evidence I can find that this was written down, but it does emphasise that salmon was eaten widely. But then, government took an interest, set up a special commission, and salmon fishing has, since then, been limited. Today it is almost a thing of the past. Which is one reason I am chasing up the country, talking to fishermen who work mostly in the rivers and very shallow estuaries and, in Scotland, in deeper water.

When man needs to take governance over something specific, he generally turns to a licensing system. These licences thus enable the authorities to restrict that activity or whatever in many ways. For commercial salmon fishing this entails limiting the fishing time, dictating the gear used, and charging a fee for the pleasure. Even back in the 1860s, there was an awareness that overfishing was a reality. Today there's definitely a lack of wild salmon, but the reasons that fishing has been first curtailed, and now banned, aren't as clear cut as they should be. Opaqueness, sometimes called politics, comes into the equation. Today, blatant lying has been made acceptable.

For salmon fishing on these two rivers, the period people are allowed to fish during has decreased over the decades and there is a charge for the licence. Until salmon fishing ceased altogether, this meant latterly three months over the summer. Generally, it was the net that was licensed and this had to adhere to certain sizes, not just in mesh but in length. Indeed, in 1948, a petition was submitted to the Minister of Agriculture and Fisheries, signed by 163 netsmen, asking for the netting season to be extended to six months because, at that time, the commercial fishers had a five-month season while the anglers could fish for seven months. Not for the first time was there animosity between commercial fishermen and the rod fishers.

The fishermen of these two rivers, as is usual among such kin countrywide, were democratic in deciding the order of fishing and gathered at various places such as the Beaver Inn in Appledore at the beginning of the season. Numbers were drawn from a hat and names of the captains of the seines from another hat, alternatively to give each boat a number. Even numbers fished the drafts on the Appledore side and the odd on the opposite, or Braunton, bank for the first two tides and then they moved across and then round in sequence. There were several drafts that could only be fished when the weather and the current were favourable.

A team of fishermen used to consist of four men and they used a stoutly built boat to row the net out. One man, the shoreman, would stay on the bank holding the shore rope, to which the end of the net was attached, while the other three stayed in the boat. The amount of rope he paid out from the shore depended on the strength of the tide and the nature of the riverbank. Two aboard the boat would then row around in a semicircle, with the third man paying out the net and then the headrope and footrope, while the shoreman began 'walking' the net in the same direction as the net (or tide) was going. Once they were close to the shore once again, this third man would throw the pole staff (a 5ft staff attached at either end of the net that stuck into the mud) into the water and leap in to hold it. The other oarsmen would secure the boat and quickly come and help him before the net was hauled in by its head and footrope, being careful not to let any fish escape by jumping over the headrope.

Although this fishing was seasonal, and therefore part-time, it could be very profitable in the days when there were plenty of salmon. In the nineteenth century, catches were good and salmon sold between 4d and 1s a pound in weight. Each crew was paid a share of the total money received from the fish dealer. A sixth went to the net, a sixth to the boat and a sixth to each crewman. Usually the 'captain' owned the boat and net; it cost at least £20 for a boat and possibly a bit more for a net and rope in the 1920s and '30s. A licence at that time cost £5 but that was often advanced by the dealer, who took a penny per fish for his trouble. Nets had to be barked to be preserved, ropes tarred and boats overhauled and painted. The year 1932 was a bonanza one in north Devon, with 6,317 salmon being netted, though, like all fishing, the next year could bring a dearth.

In more recent times a team of two can work a draft. A good example of the latter is the husband and wife team of Stephen and Sheila Taylor. Stephen has been fishing for well over sixty years, which is a long time. He's 89 as I write this, 90 coming up in a month or so. Sheila, his wife, is a few years younger and they've fished together for at least thirty-five years. I remember talking to him a year ago.

'She can haul in a 200yd net as easy as anyone,' Stephen spoke of his wife, 'and a net is lighter and easier to haul in with a fish in it.'

They still have a couple of fish huts on the Braunton side of the Taw, near where the River Caen from Braunton flows in, where they used to keep their net until their licence was removed in 2018. Access meant a mile's walk to get to them, a bit more these days as part of the marsh has collapsed. These days the huts are still full of mooring gear, anchors and buoys and things that every fisherman collects. To take away their licence was purely an act of evilness on the part of the authorities in their drive for conservation. Even the bailiff said it was a totally unreasonable act to take one away from someone who had been fishing for so long and was in his late eighties.

Stephen jokes that he had obtained his first licence almost as a mistake by the authorities. Initially licences were issued to bargemen to give them some employment during the winter when barging was impossible. Stephen decided to apply and when filling in the form wrote 'butcher' as his employment as that was what he was. But maybe his writing wasn't very

Hauling in one of the old hemp nets, Old Walls on the Braunton side of the estuary. (North Devon Maritime Museum)

clear because whoever processed the application read it as 'bargeman' and so issued the licence. That was back in the 1950s. At the time the licence cost 10s, whereas it was £350 by the time they banned all nets in the two rivers.

Often the Barnstaple fisherfolk would come down the river in their bigger boats, and some even rigged a small sail. In those days, it needed four men to handle the heavy cotton nets but once man-made materials were introduced, the lighter nets and ropes meant that only two men were able to fish one sweep. Smaller boats, such as the one Stephen still has, were ideal. The nets themselves were 200yd long and 14ft deep.

'You had to start from the shore and shoot without pause in a circle and back, and only over three parts of the river,' he told me. 'I used a 5½in mesh most of the year, though I never got a word of thanks as others used smaller mesh over much of the season. Only occasionally did I swap to the 4in mesh when the fish were running and we caught grilse.'

We talked about the highest catches:

355 in one season was the best. Sold it in my fish shop in Braunton. Some got sent to be smoked and tasted wonderful. Fourteen once in one sweep, though it might have been more as some swam out. Grilse they were. Once I caught a 38lb salmon but it was an illegal netting and I had to put it back. The man who owned my fish shop once caught a 54lb fish. We made

a good living then, had to as we had three children to bring up. There were thirty-six newsmen out on the river when I started and just three when we finished. My son had a licence too. But they kept reducing our fishing time as well as raising the cost. Used to start in April in my time but that was June by the time we finished. Up to August.

I mentioned the fact that I thought it was more about finance than conservation, the withdrawal of licences:

They didn't see that we had to make a living. It wasn't a sport like the anglers do. They catch fish and take some home and sometimes even not using flies when the fish are angry. But the netsmen and the anglers have been at loggerheads since way back. Trouble is it was a way of life and once it's gone, it's gone. We can't go back. But it has done us well. I'm 90 at Christmas and although I've had a new knee and other things, I'm fit.

'It must be all the Omega-3 you've had,' I told him as a parting shot. He laughed and agreed.

Felicity Sylvester married into a line of fishers. We met one bright Monday morning, a day after we'd both exhibited at the annual autumnal Clovelly Herring Festival, set around the picturesque harbour of the privately owned, nearby village; me smoking the herring and she exhibiting her 'Sustainable Fish Education' stand, extolling the virtues of fish while letting festival-goers taste the wonders of various dishes of marinated herring to whet their gills. This time we met first at The Seagate, a recently refurbished Appledore establishment, and talked as we sat with our cappuccinos at the low tables and chairs in the bar. It was quiet: hardly surprising for a Monday morning!

After I'd taped her 'fish talk', we left and walked across the riverside car park and down the narrow Irsha Street, with its central drainage gulley, along with pastel-coloured houses and fishermen's cottages, though I was surprised that the houses weren't of brighter colouring. In the weak November light they had none of that bright sun-filled embrace I'd sort of expected. More dim moonshine! Yet, as we ambled along, she pointed out the various homes of fishers and where the boatyards had once been, now filled in with more dwellings, until we came upon the Beaver Inn at a point midway along this strangely named street. Here, presumably on the foreshore below the wall, the fishermen once gathered at the start of the season, and then finished off their discussions over a few pints inside.

Who or what was Irsha, I wondered, but got no answer. Even Google didn't know. However, once inside and seated at a table, we were able to chat more over a drink while waiting for our ordered lunch, although our talk in lowered tones had to be so, down to the number of local fishermen who, she said, had elephant ears. Yet I wasn't here to denigrate them, although I did realise any nuances might be misunderstood.

Felicity was once a town planner until she married local fisherman Chris Sylvester in the 1980s. With long blonde hair and now into her fifties, she didn't give the appearance of being a fishwife. Maybe because she had to continue to work as well as becoming in effect what was an Appledore fishwife, she didn't fit the pattern, and she continued in the same vein even after they'd split up. Here, primed with half a pint of ale, she started telling me about her association with fishing.

'Yes, I remember some of the fishing families. There were the Eastmans, Craners, Coxes and Canns, most of whom lived in Irsha Street. Often the families got on really well, though went in different directions. Like the Coxes and the Canns, who knew each other. One family went to the pub and another to the Baptist church!'

She had a sup of beer before continuing as the chips arrived:

In Appledore people mostly fished the bottom end of the Torridge and then out to sea and across the estuary to what we call Crow Point. Names like 'Cannon', 'Greysands' and 'Pulleys' this side and 'Old Walls' and 'Sandridge' on the far side. The boats from Barnstaple would come down here and fish the Taw. My ex-husband Chris Sylvester was a salmon fisherman when he was younger and he knows a lot about the Taw. He fished with a chap called Pat Nore and they would come down salmon fishing day or night in a three-wheeler Robin Reliant and if Pat saw a rabbit run across the road on the way down to the river, he would turn round and go back and sit in the Horse and Groom pub in Braunton all day! Just because of a bloody rabbit!

Thus was the nature of his superstition. Neither did they ever go to sea on a green boat! I asked what happened to the fish when they did catch it:

We had someone catching salmon for us when Chris bought a trawler. We gave the licence for salmon to someone else and then we took all the fish and we sold it with our own fish from the trawler. So we sold river fish and sea fish and we gave them something. But I was a woman and I wouldn't be told what financial arrangements had been made. I just had to sell it. Don't ask me questions like that!

You had a licence for fishing in certain parts of the river and everyone knew their place. And if you went to the wrong place at the wrong time, boy, was that a problem. People moved round to fish drafts in a fishing time, which was quite limited round there, two or three hours at the most, and they all had to move at the same time. If you went to the wrong one at first, you'd be in trouble. It's all really complicated until you know how you fit into it. To an outsider it's almost impossible to understand. The boys above the bridge at Bideford would not be allowed to fish below the bridge but whether that was a rule from outside or a rule between them, I don't know.

'So what happened to all the fish?' I said again, chomping on a locally sourced steak sandwich:

We used to have an old deep freeze in the garage and the fish used to go in there until November, when we sent it down for smoking. I was selling wholesale and retail, as well as working and bringing up three kids, and had the shop and later a fish and chip shop. The salmon we took from the boats and sold it to the fish merchants or sold it locally. If it was a large fish, 10lb or more, we'd send it for smoking. That fish, and maybe more, would go down, wrapped up, on a bus from Bideford Quay to Newton Abbot to a very good fish shop, where it was smoked. No slicing, no vacuum packing, none of that nonsense, and it would come back, two sides together, beautifully wrapped up. And I had to go and meet the bus and there'd be mothers meeting their children and I'd be meeting the salmon on the bus. People would have it for their posh parties, hotels for their Christmas menus. I also sold naturally smoked white fish, which was unusual as, in those days, most of it was dyed. And I'm still dealing with the same smokehouse, though now it's the son I deal with. Today it's really just herring I send down there for smoking and that becomes the product I'm selling. And fresh herring of course, now, as you saw yesterday at Clovelly.

At the mention of herring and Clovelly, that was the point at which I thought I'd have to include that fish in this book. A sudden decision, though a sensible one it turned out, which then took me in the opposite direction that I had planned, and one from which I'd just come from a couple of hours earlier.

2

SWEET CLOVELLY HERRING

It's the second stop-off and already we're off course, heading in the wrong direction for Scotland; but, as I've said, it was a sudden decision. So instead of 'it's onward on our way up north', it's 'no, we're off down 'ere'! However, it would be a travesty if we didn't stop off at the previously mentioned cliff-hanging village of Clovelly, some 13 miles towards Cornwall, before heading north. It is a privately owned village that has only had three owners since the middle of the thirteenth century, nearly 800 years ago.

I've a strong affinity to Clovelly, and have had for its fishing for over the twelve years that they've been running the Clovelly Herring Festival, seeing as how I always take the Amazing Travelling Kipperhouse (an exhibition about commercial fishing) to this annual November festivity. It's a funny sort of place, with one steep street without any vehicles other than sledges pulled by donkeys when the tourists are about (when they aren't it's the locals that pull them). You'd almost think you'd fallen upon a film set, it's so quaint, though of course it has been used as such. But real people doing real jobs reside in the houses that almost trickle down the street, with their bright flowers decorating the colourful tubs outside, or around their postage-stamp gardens if they are lucky enough to have one. They say that flowers grow here 'super-abundantly' because of the unusual climate, although one blast of the north-east blizzard can kill all plants in its path. The same north-east wind has put paid to fishermen in the past too.

The herring festival is almost the last remaining of these annual celebrations around Britain's coasts, where once they were abundant. Here it's a relatively new event in the village calendar. Why? Because Clovelly has been famed for its herring for a very long time. If you look back through the annals of Clovelly history, you will find that activity around the harbour was the lifeline of this small community. The harbour, dating from the fourteenth century, was a melting pot of tradition: local fishermen from the cluster of houses lining it, coastal sailors trading from across the waters, bringing in coal for the cottages and limestone for the kiln, taking herring back over to Wales, and the odd deep-sea ship loaded with luxuries from afar, even tourists arriving from the paddle steamers from up the channel, coming for a taste of Clovelly. In summer the so-called 'long-boomers' came from Bideford, sailing trawlers that worked the offshore seas. Much of the herring was salted into barrels and sent away, keeping a cooper busy in the village making barrels. Some folk say that Clovelly herring should never be gutted fresh but left a day or two to mature their taste!

Today the tourists come by car and park atop the visitor centre, pay the entrance fee and marvel at the locals in the glass-fronted exhibition cases, for which they pay rent. Today many commute to work but in the past, work was local and often meant going to sea. Fishing throughout the year was the most profitable activity for the village: lobster and crab pots in summer, and still a bit of trawling for white fish such as cod and soles, maybe a bit of tripping too. But it was the herring that Clovelly was once famous for and which went as far as London by train from Bideford, and which many of the village's inhabitants once made a good portion of their income from. Once, so they say locally, there were as many as 100 boats going out in search of the fish and, in 1814, it is said that more than 3.6 million herring were landed by the fishermen, which was an incredible amount. In 1880, one boat alone was said to have returned to harbour with forty mease of herring, which equates to almost 25,000 fish. Locals, they say, could normally buy five fish for a penny, though at times of such huge catches the price would have dropped to more like a dozen a penny. Now, they are about 50 pence each!

I remember the first time I went fishing there. I'd just won the BBC Radio 4 Food Programme's 'Campaigner of the Year' in their Food and Farming annual awards and was recording with Sheila Dillon, the programme's presenter. Thus, I was out with local fisherman Stephen Perham, Sheila and producer Margaret Collins, shooting and hauling drift-nets not 100yd from the end of the quay.

Stephen has been herring fishing since before his early teens, out with his father who followed in the footsteps of his own father, Stephen's grandfather, and his before that. He recalls how his mum would wake him up in the early dawn, the sea outside his bedroom window, as he was expected down on the beach to help shake the herring out of the nets. He's a fifth-generation herring fisherman, his family coming to the village in the 1830s, a time when almost every man in the village went herring fishing.

However, he had to wait to fish on his own. With overfishing of herring a serious environmental disaster waiting to happen around the British coast, all herring fishing was banned under the Common Fisheries Policy of the then Common Market, until 1977, by which time Stephen's dad had died. Stephen was on his own then, in a small open boat with a few drift-nets cast over the side, working within sight of his home affronting the Clovelly harbour beach each autumn. Just him and his brother Tom still fishing in the time-honoured way. As I've said before, this book is about those who fish inshore waters close to home. For the sake of clarity, I call these 'sheltered waters', which can mean rivers, estuaries or simply close to the shore.

You could say that Stephen's a bit of a jack of all trades. You have to be these days to keep alive. He's harbourmaster, boatman, lobster potter, ex-lifeboat coxswain, harbour repair worker, tripping boat operator, unofficial bodyguard (to whom I'm not sure), rubbish putta-outa, shopkeeper (shop?) and herring fisherman, all of which ensures he's a busy soul. He herring fishes from the small 20ft picarooner, the *Little Lily*, a relatively new boat built upon the design of boat unique to Clovelly, and which was the replacement in the 1880s for the larger and more clumsy older herring boats. Because these picarooners were smaller and easier to get to sea on the rising tide, the older fishermen regarded them as cheating, or 'robbing', the word meaning 'sea robber' in Spanish, perhaps coined in the days when smuggling was almost as profitable as fishing. Although rigged with two lug sails and oars, Stephen is usually to be seen sculling the boat, standing with one oar over the transom. Thus, you can say he fishes in a totally 'green', zero-carbon way. Basically, he sculls upstream of the harbour to an area where the herring are known to spawn, shoots his drift-nets and drifts down with the current, past the harbour entrance, before hauling in his nets and returning home. Hopefully with nets full of fish, which are then shaken out once the boat is back on the beach.

Picarooners on the beach, loaded with herring, nets being shaken.

Add the recent addition of a small child to Stephen's lot, and his life is hectic but that doesn't stop him fishing. So I can simply say we'd caught up once again at the thirteenth annual Clovelly Herring Festival in November 2019, the day before talking salmon with Felicity. I wanted to start the book with salmon, which is why I'm backtracking.

My daughter Ana and I joined him on that Sunday morning. Light was just beginning to creep into the eastern sky, above the cliffs. With Clovelly facing north, the sun only spends an hour or so each morning spreading joyous rays on the harbour between about October and April, and only when there's no cloud. After two days of wind, the sea was once again flat calm and the dawn gave a bit of a chill. Stephen rowed the picarooner in the normal fashion out of the harbour and over to beyond the waterfall, a few hundred yards to the east of the village. Then, taking up a sculling oar over the stern, he proceeded to shoot the nets over the starboard side. Once the six nets were in the water, he sculled about to distract a seal lurking nearby. His brother Tom was out in a very small rowing boat, as was Vernon in another of a similar size, while Malcolm, in his fibreglass picarooner, was also nearby with limp sails set in the windless morning. The sun appeared over the hill and we were warmed as Stephen talked about the fishing:

> The second berth always has to go down below the first berth because if you were to go above them, that's shooting foul of somebody because the fish swim east to west, so you had to go the length of their nets and your nets to not shoot foul. I'm fishing six nets so I'd have to keep twelve nets away from them. You should never really go east of someone and shoot foul of someone as the fish are swimming down the shore.

The net is 18ft deep, which means we can't fish too close to the shore because of boulders. Catch one of these and it will rip the net to shreds. But the idea is to fish close to the bottom as the fish, when they see the net ahead of them, tend to swim downwards and escape under it if there is a big gap. If the water is a bit milky, then they can't see the net. This was strung out behind us, the corks fitted to the top rope every 2ft or so bobbing up and down. We checked to see if there were any herring, which there weren't, so we left them a bit longer. If there are too many herring they get pulled down under the water from the sheer weight of fish but that didn't seem to be today's problem.

Ana, my daughter, preparing kippers by hanging split herring on tenterhooks.

Two of Stephen's nets were made by his father, while the other four were his own. In the old days, of course, the nets were cotton and needed barking to stop the seawater rotting them. There were three bark houses in Clovelly, he said, one each end of the village and another midway. These tanks would have been filled with a liquid that the nets were immersed into before being hung up to dry around the harbour. The liquid was a strong tannin solution, at first oak bark boiled in water with various additives, though in the latter days before man-made fibre nets took over, cutch was used, which was imported from India in blocks resembling treacle toffee. We hauled after a while to find a dearth of herring. We chatted as we drifted back towards the harbour:

> We'd probably have caught more fish if we'd started earlier, fished in the dark rather than the light. The fish weren't playing very well anyway and probably the fish were further off because of that weather and northerly winds tend to drag things off anyway. You can always tell when a northerly wind is coming because the nets will be moving off. You can shoot up the shore but instead of drifting downwards, the nets are drifting out. The fish tend to do the same and tend to go out, instead of being pushed in with it. Always when the daylight comes the fish tend to go to deeper water even though they aren't feeding much at this time of year. Sometimes you'll see the oiliness in big spots. Herring give spots of oil while mackerel will give streaks of oil. What you are looking for is a red water because you know the fish will be playing in that, though not the stuff that comes off the waterfall!
>
> Always haul from the outside end in because the fish will be on the top of the net. You shouldn't haul from the inside to the out because the fish will be underneath the net. Inside is towards the beach. With the old cotton nets the fish would then fall out though it's not so important with

these nets because they get stuck by the gills. Always haul, too, from the end you shot last. You should do, yes. We tend to pick out the same side that we shoot but what you would normally do is you'd shake them on the other side then the fish will be under the net and you'd just shake them out. Haul on the starboard side but shake them on the port side.

Catches were poor all round we discovered on arriving back. Tommy had forty, we had three, Malcolm had ten and Vernon a few. Compare this to the haul the previous week when Stephen had 1,830 in a short time. But, as he said, the poor weather meant that even over the next few days, catches were poor. Lucky, then, that there'd been plenty from a couple of days earlier for the festival. There was no shortage when the people arrived a couple of hours later. And there wouldn't be much fear of gutting any of today's fresh herring, which might please some. Much was sold, including some 120 from the smokehouse, both kippers and bloaters, so that by 4 p.m. when they were gone, so were the people. A perfect end to the thirteenth Clovelly Herring Festival.

3

HANGING OUT
WITH HERRING

Minehead seems to be one of those places that's miles from any main transport route, it being about an hour's drive from the M5 motorway, along a winding and slow road. From Clovelly, it's a trek back along the A39 to Barnstaple and then a choice of the north Devon coast or crossing Exmoor. Both are tempting with lovely scenery and wild views. I chose the really long coastal route, through Ifracombe, with its lovely tidal harbour that was once home to several herring fishermen, past the sheltered bay of Watermouth, and stopping off for a quick walk with the dog at Lynmouth, where you can still see the remains of the fish trap on Google Earth. Continue along, with fine views over to Wales, just missing Porlock Weir with its recently restocked oyster beds of old, up, and down, up, and over and finally down to Minehead, seemingly a million miles from Clovelly.

Still, Minehead manages to attract hordes of visitors staying at the Butlin's camp close to the coast, but few of these realise just what a gem the town actually is, especially around its small, yet succinct, harbour. I'd come to experience herring fishing as I'd never seen it before.

I first met Michael Morgan and Paul Date at the Clovelly Herring Festival the previous year. That would be the twelfth one then. They are two Minehead herring fishermen, two of the last commercial fishermen in the town in fact, as they were introduced. Well, I'd already written about the local saying there of 'Herring and bread, goes the bells of Minehead' so knew what to expect. Well, I thought I did, but things don't always work out as you'd expect.

Like Clovelly, the fishermen of Minehead landed the small sweet herring that spawned there in November. In fact, it was always said that in the seventeenth and eighteenth centuries huge amounts of herring were smoked in the various smokehouses lining the quay. So there we were, on a fine autumnal afternoon, meeting up on the quay and leaning against the rails chatting about herring, waiting for the tide to ebb.

'Don't know 'bout them smokehouses,' Michael told me in his West Country drawl. 'There's no sign of any just things around the back of the houses around the quay.' He pointed over to the houses. He should know as he lives in one of them; for quite a long time, judging from his age! Then describing the other buildings, he pointed around the harbour:

That's flats, now which was warehouses before, over there, and the custom's house was there. See that line of rock, that's where the harbour wall was lower for the customs to see what's aboard. The coal was unloaded over there at the end of the quay, the coal yard is over there against the hillside, where they made coke below in those kilns. Coke for the gasworks which was there. And the church there was …

It certainly was a brief lesson in maritime Minehead. And all the time the tide seemed to be hardly ebbing!

'Be there for three o'clock,' I'd been told. Well, I was early but by four o'clock, the nets still weren't showing. But somehow it just didn't matter. Talking to Michael and Paul was much more fun and informative. It wasn't raining and quite bright for a November afternoon. Michael walked back across the road to his house to emerge after a few minutes with various photographs of boats and fishermen.

'That's my grandfather there, and father as a boy,' he said, pointing to a small child leaning against a small boat.

'He was born in the 1920s and he looks about 10 there, you'd say. So about the mid to late 1930s. Clinker-built boats with inboard engines and piles of herring nets at the stern.'

Boats built by Hinks in Appledore (Hinks again – oft mentioned in these parts!). Another photo showed the harbour with sailing boats being sculled in, probably a couple of decades earlier.

I'd unwittingly briefly met Marcus James on arrival at Minehead when I spotted him moving his 'Herring for sale' sign. He and his brother drift for

Minehead quay in the 1930s.
(Michael Morgan)

herring, as do a couple of young lads who'd just started out at the game. And Michael and Paul both had registered boats to drift with. But it wasn't drifting I was here to experience, but stake-netting for herring. Something unique.

By 4.30 p.m. the line of nets was obvious, a row of posts in a line from the road, about a hundred yards from the start of the harbour, heading seaward almost, give or take a few shakes, perpendicular to the coast. At least thirty posts Michael said, though they believed that, at one time, they might have crossed right over the bay. You can even see the posts on Google Earth! In fact, as I was to learn, you can still spot many fishing relics with the joys of Google Earth.

'Been there at least 300 years,' says Michael, though I did read that a millennium was mentioned. 'Been fished by my great-grandfather, grandfather, father and now me. We own them and no one can remove them,' he continued, 'even if they've tried.'

By 5 p.m. it was beginning to go dark and so the two then decided we'd walk down, carrying a couple of buckets in anticipation of a few herring. But the nearer we got to the nets, the more I heard Michael say 'I told you so' to Paul!

Trying not to slip on the smooth pebbles as it was pretty dark by then, and avoiding puddles, two things soon became obvious. Firstly, two buckets was somewhat an underestimate indeed. But the reason Michael was taunting Paul was he had suggested putting one net out that morning while Paul had said they'd better get three out as 'Mike's coming down and we'd best have some herring to show him'. Ha – one would have been plenty. The nets were simply overloaded with herring, and seeing them hanging in yards of netting really emphasised that glitter of the silver darlings as we shone torches. Even the light from the distant street lighting cast silvery glows like snowflakes against a darkened sky. It was mesmerising seeing the herring hanging in the nets that reached up to about 8ft at most. The poles – originally oak but now mostly chestnut from the woods above where Paul had a logging mate – towered 10ft above us and the sand and/or pebbles, knocked in about 2ft down. Ten paces apart, says Michael. The nets were 9ft deep, sometimes drift-nets cut in half as theirs were always 18ft deep. The top rope was simply wound around the top of the post and stretched across to the next post, the net hanging with lead weights along the sole rope. Fish were entrapped all over, some at the top and some tightly around the bottom where the net had wound round with the current.

We started picking the herring out and the second obvious thing was that this was going to be a long process. There were loads of herring. But, even in the dark, I soon got the hang of it. Finger under the mesh, turn the fish over so that the twine comes over the gill and flip it over the other way to clear the other gill and it's out. Some are easy, others not. At the same time, I discovered my boots had holes in but in the excitement of the haul I didn't stop. Sometimes, the odd head came away in desperation when entanglement was just too much. The two buckets soon filled so that, with the tide down, Paul returned to the harbour to fetch his 4×4, soon back with fish boxes to fill. Two hours later we had finished, and, with the nets taken down again and packed into a basket, we carried everything to the vehicle. There were 800–900 herring Michael estimated. 'Told you we should have only put one net out,' he guffawed once again!

Most of the herring had swum into the net on the ebb, caught and extracted on the side nearest to the town. Only a few were facing the other way, having been caught on the flood.

'Shows you one thing,' Michael said, holding his finger to his nose as if he'd have to kill me after letting me into the secret. I hesitated but then egged him on to tell me. 'They say to fish the herring on the flood but here, as you've seen, it's the ebb when they swim round the bay and out past the end of the harbour. They spawn here, you see.' And that had become obvious as on some occasions, the spawn had squirted out as we extracted them and you could see the spawn on the net. The ground was perfect for herring spawning, sandy as it was on the seaward end. Much of the herring caught was full of roe and when gutted there was hardly anything other than roe inside, so that those that had spawned instantly became much thinner and generally were able to swim through the net and possibly survive another day.

One thing was certain, as Michael pointed out: herring had been caught using this method for many years. He wasn't entirely sure how long the posts had been there but if you take into account the number of medieval fish weirs there are out in the bay, then probably a long time. 'Why did they place those posts where they did?' he asks. Fair point. However you look at it, they were in the perfect spot. Sheltered somewhat from the harbour – built in the seventeenth century, so did they pre-date that? – and right in the path of the herring, they remain a very efficient 'fixed engine' for their annual herring harvest. And to date, as far as I know, they remain the only example of tidal stake-nets set purely for herring. Stake-nets, as we shall see, are normally pretty intricate devices with chambers to trap fish. These are simply nets strung between posts, across the tide.

We talked a little about the fish weirs in the bay that I remember trudging through the mud to get to and surveying about 2002. With tape measure and paper, I'd drawn a few rough plans of some of them. What I didn't realise then was that they are still fished. There's a photo of Paul emptying a net that he'd placed over the mouth of one of the weirs. Dover sole, shrimps, the odd skate, he says they catch. 'They say there's seventy of them out there in the bay,' Michael told me. Now who says a fisherman is led to exaggerate on occasions?

Paul and Michael picking herring out of the net in daytime. (Sarah Date)

Paul working on his own to clear the nets. (Sarah Date)

Once we'd bumped back over the stones in Paul's 4×4 with the fish, stopped outside Michael's house right opposite the slipway and bagged up some for me to take home, I took my leave. As ever, when in the company of fishermen, I've enjoyed a refreshingly honest time, and this one was indeed no exception. Two warm characters I'd spent a wonderful afternoon and early evening with, generous in their knowledge, their experience and, of course, their time. For sure, Michael might be a dead ringer for Father Christmas, but he certainly knows his Minehead. Paul is undoubtedly the muscle of the fishery, though with the occasional reticence when it comes to lugging heavy boxes of fish around. Who can blame him? But they make a great team, as all fishing partnerships do. Some might not fully approve of the way they fish, preferring to drift for herring, but given the choice of a wet evening standing on the beach picking herring from nets, or wallowing around in a small open boat shooting and hauling nets, it's a fool's choice, some would say. Personally, I'll leave that decision for someone else to make. I'm happy to lap it all up and enjoy the working world of fishermen of the vernacular wherever they might be.

Of course, they also drift for herring in Minehead, though here they use boats with engines, unlike Stephen Perham in Clovelly. Each has to be registered, unlike one that works under sail and oar. They tended to drift on the flood, says Michael, though he thinks the ebb is best, a few hundred yards off the end of the pier. Catch 'em as they swim round the bay, as he said earlier, three or four nets is all that's needed. That easy. Ha!

Back in the van, I head east towards Watchet. I'd spent some time here with John Nash when writing *The Boats of the Somerset Levels* back in 2011. John has since passed away but I still remember him, with his enormous energy for such things, telling me about 'glatting', which must be one of the most extraordinary ways of fishing here in Britain. John gave me a DVD with some footage of a camera team and presenter, a BBC one I think, being out with two blokes, Tom and Bob, and their two mongrel dogs, Peggy and Mick. On the ledges on the foreshore just below the harbour at Watchet, they trained dogs to sniff out conger eel who reside underneath the ledges. Once the dogs sniff around a particular spot, they poke around with a glatting stick, just any old debarked stick of hazel, the fisher hoping to get the conger out. Once the conger shows itself, the dogs bite it and, with a bit of help, they get it. In the film, they found a couple of congers and the dogs helped tease them out of their holes and then picked them up in their mouths and they thrashed away, with them escaping, until caught again. With instructions such as 'bring him out boy, hold him then, good boy', the two dogs do their work that, they say, they learn as they go. No special breed of dogs. Incredible black and white footage and I still have it. Time to move on, though!

4

A MAN WITH A
HORSE WITH NO LEGS

The next stop is in Bridgwater Bay, where I was taken out to another of those extraordinary fisheries I had been lucky enough to observe, one bright blustery day in June, a bit back in years. I'd done my homework and spoken on the phone and then proceeded to find my way to the small hamlet of Stolford, only a few miles along the A39 from Minehead and then some country lanes. Maybe hamlet isn't the right word as Stolford to me seemed to be little more than an area of farmland interspersed with the occasional house. I had been directed to the Stolford Seafront car park, down Gorpit Lane, past what may be the hub of the hamlet, and atop the shingle bank of the flood defence. The second time I visited I have to say the car park was somewhat more defined than it was the first time!

Anyway, gazing out across the western end of Bridgwater Bay, it was obvious to see that the tide was quickly ebbing. I'd already called in and met Brendan Sellick in his small fish hut just along the road and, once the tide had begun to uncover the vast expanse of mudflats, Brendan's son Adrian was going to take me out to his stake-nets that lay somewhere out below the waves, which were being kicked up by the brisk north-easterly wind. It wasn't just Minehead that had nets on the beach in this part of the world, though these were very different to the Minehead herring nets.

Adrian arrived some time later in his four-wheel-drive jeep and, suitably togged up in oilskins and wellies, in I got before we trundled off, down onto the foreshore to bump and jolt along, over the boulders and stones, the track visible ahead from many years of being driven over. Then it was along a shallow stream that wound its way among the rocks, the muddy water probably almost 2ft deep. The geology was astonishing: parallel lines of clusters of broken rock shelves angled at 45 degrees running several hundred yards like soldiers relaxing while on parade. We passed through gaps where the rock had been cut away, presumably by the action of the running water. Beyond the rocks were fields of glistening mud. After fifteen minutes or so Adrian turned the vehicle around to face onshore and out we got into the world of mud. Close by was one of his mud-horses, a wooden structure some 6ft long and 3ft wide with elm boards curving up at the front to form a wide skid (no legs, see) and which was weighted down with a number of large stones to prevent the tide washing it away. This was the vehicle to transport fish across the vast mudflats and it looked almost prehistoric with its worn timbers and mud encrustation, though Adrian told me it was only 3 years old. He then instructed me to walk along one of the seaweed-covered stony banks, across some mud and to follow the line of stakes to where his nets were, visible several hundred yards further out. He began to lift off the stones to free the mud-horse for he was going to take the direct route to his nets, pushing his horse across what looked like a vast sea of thick mud. Off he went, pushing smoothly as he trampled through the mud with an amazing confidence. Then he stopped for a breather, standing right on top of a rock of which he knew the exact position. He must have, as it still lurked several inches below the mud. To miss it would be to stand in several feet of mud, which could be sticky to say the least. Or even lead to a sticky end. Then, just as suddenly, he was off again and across to more solid ground, a constant pace, each footstep only momentarily staying on the mud.

Meanwhile, I had a good trudge through mud that flowed over the tops of my wellies and which almost succeeded in making me fall over when attempting to take a photograph. Yet I, too, arrived on the solid ground where the nets were. These were strung to a series of posts perpendicular to the shoreline and held up by posts into the mud which were some 6ft high. I should have followed Adrian's example, though, I thought as I sloshed around with boots full of water, for he was wearing shorts and trainers, though he did admit that he didn't always.

'I persevere as long as I can in shorts but in winter it gets too cold.'

'You mean you come out here in winter?' I replied in surprise.

'All year round when the tides are right. Maybe on the neaps I have a few days off. Sometimes the weather is too bad, but, yes, most days I'm here. It used to be me and Father but now it's just me.'

Postcard showing Brendan pushing his mud-horse.

We inspect the nets. Firstly there is a line of twenty-eight shrimp nets, nets with a square mouth several feet across and which are funnel shaped to decrease to the cod end some 6ft downstream. Each is untied, the contents emptied into a sieve before the net is carefully tied up again. The catch is sorted – the small fish, weed and bits of rubbish are extracted – after which it is added to the slowly filling basket. This catch consists of shrimps, dabs, Dover sole, whiting, the odd mullet, dogfish and skate. The latter was expertly sliced to separate the edible wings. You can certainly tell an experienced fisherman by the way he cuts a skate. Then he cleared the few stake-nets, which held the odd fish, before we trudged another few hundred yards through mud and water out to a further line of stake-nets set out by the low water mark. Here were two more skate, mullet, dogfish and a couple of decent-sized Dover sole.

'Do you come out here every time?' I asked.

'Normally but there wasn't much yesterday so I thought that I might hang them up today if there was nothing. It saves the tide ripping them. This wind brings a lot of weed down off the beaches too. But there's not much time out here. By the time we get back to the jeep this lot will be under water.'

'Have you ever been caught out?'

47

'Once. Then you have to walk right round there.' He pointed to a line of stakes further upstream on what looked like solid rock. 'It's much further but I put those stakes there just in case.'

As we sloshed our way back, my boot still full of God knows what, the dangers out here were obvious. My feet seemed encased in concrete with some bits of glass embedded into me. This obviously wasn't a place to mess about in, especially where the mud was many feet deep in places. Once we arrived back at the mud-horse, Adrian loaded up his baskets and, while I retraced my steps through the stream and mud, he pushed his contraption straight across the mud back to the jeep. By leaning upon the framework, he can push it forward without his feet sinking too far into the mud. However it did look hard work.

'Not half as hard as carrying the fish in a basket though,' said Adrian when I mentioned that. He had a point.

'Years ago we had to push it right to the foreshore. There was no driving the jeep down here,' he told me as he unloaded the baskets and weighted down the mud-horse with the large boulders again. 'It was all mud, no rock. Until they built the power station there. Changed the nature of the beach then, they did, and almost destroyed the fishing too. They suck in more fish than I can dream of in one day. They have a filter in their cooling water intake and they used to put the fish in a skip round the back. Stank, it did, at times. Now they don't but they seem to hide it.'

The power station in question was the nuclear one at Hinckley Point, of which there were two, although one has since been decommissioned. But today the French company EDF was building another reactor, Hinckley C, financed with Chinese money, a deal meant to contribute to keeping the lights on when all the other nuclear generators shut down. Police patrols had become almost constant, a sinister development, especially given the formation of a dedicated squad of armed police for this sole purpose.

The power station had, over many years, been responsible for a reduction in the catches the Sellicks collected. In the late 1950s, they were landing 4,500kg of fish a year but, after Hinckley A & B were built, this had reduced to about 1,000kg by about 2010. Now he's lucky to get a fiver's worth since they've been building the new station, along with a new pier to bring in components directly by sea. The underwater vibrations from the drilling for this and the cooling system play havoc with the fish. But, of course, neither EDF nor the British government care about the employment of one man and have certainly given the Sellicks no support whatsoever, so Adrian is slowly being squeezed out. The normal excuse from EDF's spokesman has been that all company activities are controlled and regulated by statutory bodies to ensure the environment and public are protected and that there is no threat from the dredging. What a load of crap!

We soon arrived back at Brendan's fish hut, where he was preparing to process the catch. Adrian went off for a deserved shower while I was determined to ask Brendan more about the traditions of this fishery, which I presumed was an age-old method:

Well, I started when I was 14, when I was a kid. My dad was here all his life and I carried on from him. I've done it for sixty odd years, my dad did it the same, his father did it the same and the grandfather before him started it in 1820 or something. He was a stonemason, he picked up with a local fisher girl and she persuaded him there was no money in fishing and that's how we came here, away from the coast a bit. Of course there were a lot of families doing it in those days. Any amount. When I started there were three or four families, a dozen blokes, nine brothers and that. Twenty-five years ago there were two families. All had their own patch and a licence. That's only a few quid today but it gives us reassurance.

He paused for a few seconds before continuing:

Yes, non-stop for sixty odd years, I can't believe it. Adrian went out when he was 8 or so. Some blokes from Burnham did it using a boat to get out to their nets. I remember one father and son one Sunday whilst we were out. They disappeared whilst out trammelling in their punt. Went in they did and weren't found for three weeks. Probably one got caught in the net when it went over and the other went in to save him. Lost their knife, they must have. Yes, many have lost their lives out here where all sorts of things can happen.

I'd heard that there were mud-fishermen over on the Welsh coast so I asked him about it:

Back in the early 1800s there were too many here so a couple of them moved over there. Cardiff Bay. One of them had a shop in Splott and he did the market and also had a barrow. Then my uncle George Sellick did it over in Cardiff. They went from here to Cardiff, his father did it. It finished up in 1939 at the outbreak of war. Bombing times. We were stopped for fishing for one year during the war. One year. That's all.

A family of customers arrived to buy fish. Eventually they decided on a dozen dabs – dabbies they called them – which cost £9.50, which seemed very reasonable. Two skate wings had cost an earlier customer £5.

'People phone up. That's how we want it. Some of the time we had to stop people as we had nothing. So's they phone and ask what time to come. Most of them do.'

I asked what the catch generally is. Was today's typical?

Shrimps are coming in now. These are a bit small but they'll start coming now and once they start coming they'll come thick and get bigger. Brown shrimps. Skate, Dover sole, whiting, bass in the summer. No bass today but some days there's half a dozen. A lot of mullet, flounders, dabs. Doggies I skin and sell for £1.50. We used to go all over the place. Bristol, Cardiff, Weston, London even. Took it to Bridgwater to put on the train. Twenty years ago or more. Went there by horse and cart at one time. Then I used to drive over to Weston twice a week to deliver to about twelve fish shops. None there now, I think. One in Burnham maybe. Lloyds the fish merchants in Whiteladies Road in Bristol. He used to ring up and he'd be here in an hour. Different people would come and pick it up. Couple of hotels ring up and take a few mullet or a bass. Old folk used to come and buy shrimps to sell around the pubs. It used to supplement their pension. They can't do that now. In autumn that's the best time. A lot of skate, cod, whiting and coming into winter sprats and whitebait. Little baby whitebait. Cod, cod and more cod in winter. Bass very early this year, earlier than anywhere else. Think they come up channel to spawn. Yes, it was quite an industry years ago. We had seventy or a hundred shrimp nets and much more gill-netting. Now we are the last and just manage to survive.

'What about salmon?' I had to ask. He laughed.

'Used to very occasionally but not these days. There's none out there.'

'Herring?'

Nets facing the ebb, set some distance out from dry land. The mud-horse sits waiting!

'Again the odd one gets caught but almost as seldom as salmon used to. Possibly a bit more these days, though I can't remember when I got my last herring.'

Adrian had already told me that he couldn't have more nets because he wouldn't have enough time to empty them. Two hours out on the mud is a maximum and sometimes it's less. When the two of them fished they could service more nets, but now that he's alone he can't and his own sons aren't keen to carry on. To take on someone else would mean having to pay out wages, which the business wouldn't support:

I make the nets in winter as well. I made a lot this winter. Hell of a lot. Twenty over the last couple of months. You can't do them in five minutes as there's quite a bit of work in them shrimp nets. They have to be done properly or they wouldn't last five minutes. If you had to buy them they'd charge quite a lot of money. It takes several hours to make one of them.

Brendan had lit the shrimp boiler in preparation to boil the shrimps that he had been washing while chatting:

These little shrimps are very popular. Brown shrimps. In a month's time when they get a bit bigger and the quality's a bit better, they sell very well. Yes, very well. We get a few pink shrimps when the wind goes to the west and blows them out of the rocks. And a few prawns. But 90 per cent is brown shrimp. See the white shrimp there. A sort of albino shrimp. We call them whiteshanks. You can eat them. Seems they are peculiar to the Bristol Channel and they used to catch a lot of them down at Watchet.

The first batch of shrimps went into the boiler, where they stayed for a few minutes before Brendan tipped them into a large round shallow basket he called a 'reap'. They used to have two dozen reaps full of shrimps while today he'd be lucky to fill half of one. We talked about the loss of mud, which he put down to the actual structure of the power station that, he thought, caused eddies in the prevailing wind that moved the stream. The mud has undoubtedly gone down several feet in depth and, as both Sellicks pointed out, Mother Nature usually takes generations to make such drastic changes; they had only noticed the mud removal since the power station was built. However, as is always the case, the lack of precise data means that British Energy or anyone associated with them simply deny the facts. To them it's either a natural phenomenon or simply a coincidence. Then the talk moved onto supermarkets, as it generally does, and how fish shops had since disappeared. Why, I wondered, is it always the large conglomerates that destroyed the livelihoods of those following the traditional ways that had otherwise survived generations before they came along. By the time we'd finished, Brendan had boiled the entire catch, which didn't amount to much more than 15lb. Then, just as I was leaving, he showed me the eel tank outside, alongside the shed. Eel catches, though, were poor again, largely because a ton of elvers had been caught from the River Parrett and had been shipped straight to Japan. Not surprisingly, the eel population had shrunk drastically.

As I left I wondered what will happen when Adrian packs up. Are grants available? As Brendan says, 'They give a man thirty-thousand a year and a new Land Rover just to count the birds or to see how many mushrooms are going down the field and to record everything. So why not to keep the mud-horse going?'

Exactly. Why not provide some assistance to keep these traditions going? As he spoke, I suddenly pictured the film crew that Adrian told me he once took out. I could see them with their gear trudging through the mud. They were Americans, he said. I think of the film people I'd come across with their pretences and arrogance and somehow couldn't see them plodding out to film Adrian at work. Thankfully these did, but in time their footage might be the only evidence that mud-horse fishing ever existed. Luckily Adrian, he told me, loves it out on the mud, the day-dreaming and peace away from the pressure of society. True, he has another job working nights in the local yoghurt factory and I have to wonder how and when he sleeps sometimes. But to both him and Brendan they have something worth hanging on to, in their eyes, something unique in British waters and something that not only brings in an income, however small, but remains a tangible link to their family's past.

I phoned for an update prior to submitting this manuscript. I spoke to Brendan, who'd be 90 in a few years. Adrian was still at it, though that week the weather had stopped him. The only fish they had at the shop was frozen. Shrimps had stopped at Christmas but hopefully there'd be white fish before April. They were still living at Mudhorse Cottage, which was reassuring. 'Don't think we'll be at it much longer,' Brendan said as a closing comment when I mentioned the new power station. One can only hope that this does not become yet another tradition faded into the memory of the few that care. But I'm not holding my breath.

5

PARRETT FISHING

Bridgwater, a stone's throw from Stolford, is a sort of mucky place. Maybe that's unfair but it is the impression I get from various visits over the years. You could say it's not my favourite place, which will seem unreasonable to many of the town's inhabitants, though surely some agree. Part of the muckiness – maybe most of it – comes, literally, from the River Parrett. It divides the town because it is regarded as one of the most dangerous rivers in Britain due to the large tidal difference and the fact that it is immensely muddy with all the silt it carries down. Indeed, it is said that more than a hundred folk drowned in the river in the nineteenth century because of the density of the water, and that ships too were caught out and subsequently sunk in the mire. If you fell in, it sucked you down a bit like quicksand. Such an unremarkable river with remarkable effects!

If the river is so silted you might be surprised to hear that Bridgwater has been a major port back since at least the fourteenth century, when pilots were based at Combwich Pill and ships sailed up the murky waters to the town bridge where goods were unloaded into barges for passage upriver and inland. Half a mile north of the bridge, on the west bank, is the entry/exit of the 14.5-mile Bridgwater and Taunton Canal, which dates from its opening in 1827 and was built to connect Taunton to Bristol by water.

Bridgwater also built ships, fine wooden vessels, all 167 of them, with the last one being the West Country trading ketch *Irene*, which is still sailing as a charter vessel. The stonework of East Quay can still be seen alongside the tree-lined road although, with industrial and retail sites across the road, it is hard to visualise any of the shipbuilding these days. However, trace this road south to the town bridge and the road becomes 'Salmon Parade', which conjures up entirely different images.

Today the river is better known for its eels than salmon, with thousands of the slippery fellows heading into the river and upstream in the early parts of the year. But salmon were once also swimming upstream to breed and where there's salmon, there's salmon fishing. Well, at least there was. Today, along with the near horizontal rain and the noise of the traffic, not to mention avoiding the puddles, I'm finding it hard to visualise anything but desperation and dankness.

One well-known person from hereabout I wish I'd met was River Parrett farmer/fisherman Bob Thorne. Bob came from a family of river people who maintained ranks of putchers at Black Rock, just inside the River Parrett, before him, as well as holding fishing licences. But sadly I'm a few years too late.

Born and bred by the riverbank, he lived in the thatched sixteenth-century family house near Pawlett, approximately one mile from Black Rock, almost continuously up to his death in 2013 at the age of 88. As a child he was sometimes taken to Sunday school by boat, such was the life on these levels. He was called up during the Second World War and became a medic. Afterwards he returned to what the family had been doing for generations – digging, gardening, fishing, shooting rabbits and pigeons, tilling the fields – a bit of everything in fact, including brief periods of jobbing at times, and had been doing the same in his later years. Even into his eighties, he tended his apple trees, pear trees, cherry trees, plum trees, walnut trees, made his own cider and grew his own vegetables. He never married and always claimed to have had no interest in women.

Just before I continue on Bob's story, a wee explanation is necessary on traditional fish baskets. Both putchers, and the larger versions called putts, are conical-shaped baskets made from willow. Putchers were the smaller baskets, some 5 or 6ft long, which were arranged in what are called ranks, up to six putchers high in some places, and many alongside, though Bob's were only two high. Putts come in three separate sections – the kype, butt and forewheel – which, when fixed together, meant a fishing trap up to 15ft in length. The kype was the mouth with an opening 6ft in diameter, while the smaller forewheel was where the fish became trapped as it swam in. Although extinct these days, several would be placed on the foreshore along-side each other, though during the off-season the mouth of the putt had to be closed off. We will learn much more of both of these fishing baskets in a later chapter, though it is a fact that, before 1866, there were more than 11,000 putchers on the River Severn alone with hundreds more on the other rivers flowing into the Bristol Channel such as the Parrett.

Eddie and Cecil Reasons dipping for salmon in the late 1980s. (John Nash)

Back, then, to Bob who had started fishing with his own flatner, the local Somerset flat-bottomed boat, in 1949. One of the local boatbuilders, Bill Pococke, renowned for building good boats, had built it for him, though Bob had to run his dad's coal business that year so he didn't fish with it until the next. He caught his first salmon in May 1950, a seven-pounder. Not remarkable again, but a good start!

Since then he'd had licences throughout his fishing career for both putchers and netting. He had to cross several fields, climb over stiles and trudge along the river twice a day to check his ranks. When the tide was bad, he'd either tend his stake-nets or go dip netting or pitching. Then, when the ranks had to be dismantled at the end of the fishing season, he had to carry each and every putcher back to his storage shed on his own, a laborious task in anyone's mind. Then, before the start of the season, he had to carry them all back again. Having somewhere in the region of 250 putchers in several ranks (fifty putchers to a rank was the maximum), of which it is only possible to carry a few at a time, necessitated endless journeys. He had a small fish shed by the river, the first of which was supposedly built by Joseph Reasons in 1852. In this he kept most of his fishing gear and, until recently anyway, his last flatner, a glass-fibre boat, still sat there rotting away, though the shed itself blew away several years ago.

The 'Hang' was a large (200yd-long) type of stake-net very close to the mouth of the River Parrett, which is marked on maps dating back to 1800 but is deemed much older; though it has not been used for many years. Putchers and putts were also constructed here.

Pitching is another way that Bob Thorne, and others, fished salmon on the Parrett and is perhaps unique as I've not come across it anywhere else, although there are similarities to both stop-netting and compass-netting, methods we shall meet in time. The difference is mainly in the way the vessels are utilised.

In the case of 'pitching', the fishermen would moor two flatners alongside each other off the rocks with anchors bow and stern so that the boats lay across the stream. The net was fixed to a 'V'-shaped affair, about 20ft across the mouth on 15ft poles. This was lowered into the water so that the bag of the net trailed beneath both boats and it was held down by the fisherman who stood in the second boat, even though the net was balanced against the gunwale of the first boat. The presence of a fish would be felt through 'feeling string', which were attached to parts of the net. Bob Thorne once gave a beautiful description, which I quote:

What we used to do see, you 'ad your two flatners, used to go out off the rocks on neap tides, 9ft and under, anything over 9ft it'd run too 'ard. 'Course on the spring tides, salmon used to go upriver. Right upriver, well they didn't never used to get back. They used to stop upriver 'cos there was plenty of water. Then when as the neap tides do come so the salmon would work back, as water gets scarce and like in the summer very often there wasn't enough water for 'em to go on up in the fresh, see, they'd 'ave to 'ang about river till there was enough water for 'em to go on up fresh, see? Or perhaps for the next spring tide when twas 'igh 'nough for 'em to go on up. Or they used to work back river. Well then what we used to do is on the flood tide to go out Black Rock, put your two boats, er, you'd have your first boat with two anchors on, one bow, one stern, chuck they out, chuck the two anchors out see, so as you 'ad 'er broadside to the tide. Then you'd come in with your other boat, tie he to'n and put your two dollies, used to have these two sacks with straw in see, put the two dollies one each side the thole pins, you know so she couldn't chaff, see? Well then you had your net which was two long poles, 15ft poles and with a 28/30ft headline for it. And then, soon as tide turns, 'e'd go out over side of the boat see, and drop down on the bottom because in the river there out on the Rock at low tide there'd be about 2ft of water. Well then at the end of th'our on a neap tide you'd 'av p'haps six, seven foot of water. You'd fish for a hour then you'd chuck in because any fish about would be gone on back, see? And that's 'ow we used to do it, see?

The net was out over the first boat, and you was stood in the second boat, like, the boat behind. Then when you did weigh your net down, generally two of 'ee there see, and one, like the one, er, 'ad a axe in 'is 'and so that if anythink went wrong, you know like the boat started to tip or anything, and you couldn't release the net, 'e'd just chop the anchor line and the boat would swing round, bow on to the tide, see?

Two straight pole with the net in between. Plenty of bosom, see, and the bosom used to go straight back under the boat. Well then you 'ad your finger in the mesh, see, and soon's you felt a tug you knew something 'ad gone in the net. Well sometimes 'twould be a flatfish, could be a big eel, anything. But you know like I've gone out there three or four days with th'uncle and 'ad a fish or nothing. 'Ad one or two flatfish and that, nothing. Then all of a sudden you go out there, catch the right tide when the fish 'ave gone back down and 'ave perhaps two or three fish there, in thick hour, see? That's 'ow it used to work.

There were once some twenty licences issued to permit fishing in the river, though catches declined after the 1930s, due partly to pollution.

Bill Pococke, known as 'Pokey', was another local character, renowned for fishing with a dip net from the riverbank in Salmon Parade, near where he lived. It was legal, according to by-laws, to be able to fish from the riverbank or even the pavement, in this way, though not on a Sunday when it was not, as long as a licence was held. To do this, a dip net was used, but this had to be used in a particular way in that it wasn't dipped, as it was from a boat, but whacked down instead. Bill Pococke, dressed in his trilby hat, was, it appears, particularly good at doing that!

Dip nets were normally used from a flatner to catch salmon as they swam upriver with the incoming tide on the Parrett and were confined to the lower part of the river between Dunhill and Bridgwater. Each net was licensed and the licence number had to be displayed upon the net. The fishermen rowed with the tide and noted the spot where the fish came up to clear his gills. As mentioned previously, the Parrett was, still is, a very muddy river, and the salmon couldn't travel more than about 25yd without clearing. Having noted the spot, he let his oars go to be caught in the thole pins, ran to the bows of his flatner and dipped the fish next time it appeared. The fish was killed with a bash on the head from a 'priest' or 'killing stick'. It sounds easy, and has been described as a 'sport', but it was in fact hard work, and a keen sense of balance was required. Some of the boats used for this type of fishing were not fitted with thwarts, but had a central box that served as a rowing thwart and storage. The obvious advantage is a clear run forward. The traditional pattern of dip net was like a large 'Y' with a 6ft headline (the legal maximum) and a long handle. A more conventional net superseded these cumbersome nets, rather like an overgrown butterfly net, sometime after 1945. They only survived that long as Somerset fishermen were, by nature, very conservative. 'If 'twas good enough for Father, 'tis good enough for me,' was their attitude, similar to that of many other fishing communities up and down the west side of the country.

Both Bob Thorne and Bill Pococke are long gone, as are the salmon, so we are told. But their spirit carries on, and I'm sure I saw, among the drizzle dripping down from my forehead, a hazy figure, trilby atop, lean over the railings of the town bridge, gazing with intensity upstream, alongside Salmon Parade. Maybe I was mistaken. Cold, wet and hungry, it's easy to believe anything! But that ripple on the river, that unmistakable splat, that shadow beneath the old bridge, that was no dream.

6

THE WESTON BAY SHRIMPERS

Weston-super-Mare is nothing like Bridgwater indeed. For one thing it's on the coast, with its two piers jutting out into the murky depths of the tidal Bristol Channel, one of which is derelict. It's a Mecca for those with nothing much else to do: plenty of amusement arcades and touristy shops to satiate the appetite and then rob you of your well-earned income. Day and staying-over trippers have been coming here in hordes for at least a century, though these days those that stay overnight have decreased as day trippers have increased. On the plus side, there's a wonderful long sandy beach and, with a huge tidal range, the sea goes out a long way, which can cause consternation among lifeguards and the local lifeboat. Traditionally, the small Weston flatners had been the local fishing vessel, working close inshore, more often with Victorian trippers than fishermen, but these have all but gone and modern engined replacements are few and far between.

I went down to Weston-super-Mare to meet the bay shrimpers at the end of their season. They don't need boats as they fish their nets just a few yards away from the shore. It was early December, I think, cold, wet and grey. And before you ask: yes, I know there's no salmon here, but after talking to the Sellicks, I thought I might just make a refreshing change and experience a bit of shrimping. They catch the occasional herring too, so we might be lucky this time! And so this book's fish portfolio increases to three!

I'd first found Richard Reynolds through Facebook and contacted him. This was to become a trend, this social media thing! Then, with a couple of weeks of easterly winds following our initial contact, I had to wait. Winds from the north or east stream the stall nets in the opposite direction and can damage the poles. With south-westerlies being the obvious prevailing winds,

I had to be patient, something I'm not good at. I had the bit between my teeth and was eager to see what it was about before the season ended. Happily, before long, things were looking good and I had the green light. Thus, when I did arrive with instructions where to meet, it was easy to spot the fishermen out of the crowd; they were the ones in yellow!

They'd had a couple of days of strong south-westerlies and weed had been problematic in that it was getting caught up in the net and blocking the cod end so that the force of the tide wasn't able to push prawns and sprats into it, allowing them to escape through the net. Although this in the main made clearing difficult, it, in turn, reduced the amount of catch. Having said that, the day I was there the catch was the poorest of the season, primarily because of the weed, though the day before, even with weed present, there'd been 7lb of shrimps and plenty of sprats.

The Weston Bay shrimpers, as they are collectively known, are all either retired or work elsewhere. For example, Richard has a full-time job.

'I've a garage, so that means I can leave if we're not too busy, leave the man who works for me for an hour or two. My dad is retired so he's always able to come down.'

There were four of them on this occasion, even if there are more: Richard, his father Geoff 'Norway' Reynolds (they say he always gets the big cod!), Geoff Rich and Dave Milverton. They all say they simply do it for the social connections they've made, and to keep the traditions alive. In other words, much in the same way as many of the fishermen I've met over the years, and

Postcard view of Weston Bay shrimping grounds.

whom we shall meet over the pages of this book, they spend their time (and money) purely because they believe it is really important to keep these rural ways of life, that have passed down through generations, alive and reckon that once they stop they will disappear for good. The one common problem they all have is encouraging the young of the next generation to get involved. Here there's more continuity in the distance as Richard says his 12-year-old son Jack is biting at the bit, and comes down whenever he can and contributes. He's also learned to mend nets and will bring his net needle along and make repairs on the beach.

The other requirement is not to demand anything in return except the occasional fish. Richard says he's not keen on eating fish, so doesn't mind not getting any benefit at all from the work. It's enough to see Jack learning. That's just how it is. The nets, you see, are owned by David Davis, who we meet by the RNLI gift shop and who supplies the trolley, baskets and sieves from his car. These we took down to the nets when the tide had ebbed enough. They are then returned to him and he also takes the catch as well as the tools. The shrimps are soon boiled and they go back to the fishermen or to friends as they aren't able to sell the catch. If they did, they'd become commercial and beholden to all the regulations.

The nets themselves are situated by the Birnbeck Pier, the structure at the north end of town that leads over to Birnbeck Island, known locally as the Old Pier. That in itself is an intriguing structure, built in the 1860s as the only pier to an island that became the Victorian equivalent of a modern-day theme park. The old lifeboat shed stands forlornly, the lower part of its launching ramp destroyed by the weather. Because of the derelict state of the pier, and the buildings on the island, judging by their desolate, eerie appearance, it's been closed and sealed since 1994.

The fishermen have some nine nets, rectangular in their mouth and about 8ft long leading to the cod end, strung between poles set about 8ft apart and 6ft high, facing the flood tide, though only six were set when I was there because of the weather having bent over some of the poles, caused by the weight of weed on them during the strong winds and spring tides. The beach is solid rock and huge stones, thus erecting these 2in diameter tubular scaffold poles is itself an art and when one snaps, it's a job to replace it. But they do, they have to, in the time-honoured way of brute force!

Likewise, the fishing itself isn't rocket science: you untie the cod end when the water has ebbed enough and tip the contents into a basket, just

The furthest out nets after a good southwesterly gale.

like Adrian Sellick does. If there's loads of weed, then mostly this is done the other way round by pouring it through the mouth of the net into the basket. Then the contents are sorted, weed cast out, sieved when possible and shrimps and sprats put aside. We had a few live tiny sole, which were put back into the sea, one herring that had been attacked by crabs, a few crabs themselves, and other small fry. The odd weevil fish was something of which to be aware. Nets are then checked and retied, or taken down and returned ashore if in need of repair. Orderly and efficient.

Richard told me about other stall nets around as well as the trammel nets some few hundred yards west of where we were. There were once shrimp nets in Sand Bay, and even today a single one is still sometimes set and operated by a lone man whom Richard thought was Polish. Others were once to be found around the headland of Sand Point to the north of Sand Bay, where I later went stomping off to look for the shrimp boiling hut atop the cliff, although to no avail. But as I stood atop, in a field full of sheep, viewing down onto the expanse of glistening mud and sea that was the Bristol Channel, where various visible poles in the mud could once have held nets,

I thought how this water was generally considered as being devoid of sea life. On the contrary, it has a huge tidal range that needs a strong and powerful current to fill and flush it twice a day. Surely that in itself is a good supplier of energy to bring in fish? The fishermen tell stories of turbot and plaice, of sole and mullet, the occasional skate, big cod and an abundance of shrimps. Add in the sprats and herring and maybe, because the Bristol Channel isn't the domain of fishing craft, the fish have a chance of survival and, in turn, are able to enrich these waters, as well as themselves. But only as long as the traditional fishermen are able to thrive with the restriction on trawling that keeps the big boats out.

PART 2

RIVERS SEVERN AND WYE AND INTO WALES

7

PUTCHERING AND PUTTING IN THE RIVER SEVERN

It wasn't merely the River Parrett and its locality that had its putchers and putts, but those in the River Severn far outnumbered anything set elsewhere in England. Although evidence of wicker fish traps have emerged from the muddy bed of the river over time, most have not been dated to anything beyond the seventeenth century, a time when a reference was made in a memorial song recorded in 1663 of two fixed engines operating 'between the Hill and the Pile'. These were presumed to be Hill Farm (fairly obvious due to the lack of hills hereabout) and the Mireland Pill Reen or 'Monksditch' near the small village of Goldcliff, which is a fairly isolated spot on the Gwent Levels, just east of the River Usk's estuary. The nearby ruin of the Benedictine priory gives a clue to the siting of weirs (and the name Monksditch) just to the east of the remains, under a small promontory, while the Ordnance Survey Explorer map still shows a line of where one of the weirs was, at what is in essence, the 'Gold Cliff'. However, on closer inspection, I failed to find any gold, though the line of posts in the mud betrays the position of an old putcher rank. There's little sign of any cliff either as the riverbank is built up. However, it is believed that weir fishing here pre-dates the medieval period. This was also one of the last working putcher ranks in Wales. If not the last! There were originally three ranks – the 'Flood', the 'Ebb' and the 'Putt', and together these three were able to carry a total of 2,327 baskets. Imagine having to replace that lot one winter.

Upstream of Goldcliff there were various ranks laid out into the river, and if you count all those between say Beachley – where the Wye River Authority once ran a rank of 500 putchers – and say Gloucester, then you'd be amazed that a fish managed to swim upstream, which was one reason limits were put on the salmon fishing in 1861 and 1865. That was after the government set up a Royal Commission to enquire into the salmon fisheries of England and Wales in 1860. The number of allowable putchers was determined by a Certificate of Privilege: if someone could prove operation of a weir that pre-dated 1861 then they obtained their certificate. No new putcher weirs were allowed. One statistic tells us that, in 1863, 11,200 putchers were in place on the Severn.

In the lower reaches of the river, some 25 or so miles up the Bristol Channel from Weston, lie the two Severn crossings. Before the first bridge was opened in 1966, two ferries plied the crossing. At Aust, a ferry crossing dates from Roman times and survived right up to the day before the bridge was opened in September of that year. Cars were introduced in 1926 and this became known as the Old Passage crossing. Another ferry rivalled this one at New Passage, a few miles downstream, especially when terminals for the railway were built on either side of the river – at Pilning and Sudbrooke – although this ceased working abruptly in 1886 when the Severn Rail Tunnel was opened.

Walk along the road, past the remains of the ferry terminal at Old Passage, Aust, and along to the first Severn Bridge, and it's easy to spot the remains of a rank of putchers down in the riverbed at low water. A line of posts betrays the place, just as at Goldcliff. These putchers faced downstream, whereas the majority faced upstream, catching salmon on the ebb. There were 640 putchers here as well as 18yd of hedging, in the form of posts and interlaced withies, which helped to 'lead' the fish into the traps. This fishery, known as the Salmon Weir, continued operation to 1996.

It was fished by Norman Knapp, a farmer from between Almondsbury and Tockington, who had fished it from the early 1950s. Originally given the right to fish 640 putchers in 1866, this had reduced to 600 by the early 1960s, a time that Steve Tazewell worked for Knapp at weekends. He worked both upon the farm and for the fishery, depending on the time of year and Knapp's own workload.

Steve was apprenticed in a garage, learning the motor trade, which he's still, just about, in, although his son runs things these days. Is there something about garages and fish, I wondered, recalling that Richard Reynolds also had a garage.

Steve says he remembers the construction of the Severn Bridge well. We talked about the rank of putchers, mounted on spars some 20ft long. Three tiers he thinks, although he says his memory of those days is fading.

Norman's dad used to work in the ferry office when the ferry was working and then, when the bridge was built and the ferry disbanded, he worked on the tolls. Presumably, this is how Norman ended up with these ranks, even though he lived some 7 miles away:

I remember watching Norman making the putchers with a framework. No, I never made any. He got his willows from the banks of the river where they'd been put in to stabilise the bank. When I was a kid, say if you were at Almondsbury and faced the Severn, all that there was was wetlands.

Even today, looking at the Ordnance Survey, there's various Rhines here – Bunsham Rhine, Moor Rhine, Tockington New Rhine for example – and the area is also interspersed with withy beds such as Gussy's Withy Bed and Old Withy Bed:

See I was working weekends for him, Saturday and Sunday, as we lived opposite him. He'd get teenagers like me to help him. And he'd pay well, thirty shillings for two days. I started on two-pound-fifty for a week's work as an apprentice so thirty shillings for the weekend was good money. Most weekends during the season that started 15th April and lasted until 15th August. But it gave him a bit of free time during the weekends. He'd milk and not have to fish. But he'd do it all during the week, milking and fishing, so we gave him some weekend rest. And he could afford it as the fish made good money back then.

So I asked him what the job of emptying the putchers actually involved, given he'd be down there on his own. I'm not sure how today's health and safety inspectors would react to that!

Well, first of all I used to have to cycle down but sometimes his wife would drive us down if she was around and come back a few hours later, and eventually I got my motorbike licence then he'd let me use his motorbike. That was fun riding his motorbike down there. The worse bit, and I used to fear this, was there being a fish under the water. See, I used to have to stand on a rail, emptying the putchers. You had to get there as soon as the water went down. Once they uncover the crows are there. You cannot be there as they uncover so you walk out with the tide, all the times. Sometimes you'd go down there and it would be pure rock, you could walk out with your boots and you'd come back to the motorbike and your boots were as clean as when you went. Then the next day you could go out there it was as deep as that with mud.

We talked about the tide and the remains and then how you actually got the fish out:

The water was really murky and you couldn't see much in it. You'd walk along the rail with 15ft of water swirling around below you at the far end when sometimes the tide didn't go out far enough. Maybe the bottom two putchers at the bottom end still flooded. So you had a stick, which was always jammed in the putchers, and you'd go down to the nose of each putcher, and you'd have to find each one, and go into the nose to make sure there was no fish in. If you left a fish in, it went white and you couldn't sell it to the hotels. So if you had a fish in, you had to climb round to the other side without a rail, and the only way you could do it was to put a hessian sack, as we all had hessian sacks, with your hand in it because if you tried to grab a fish weighing 20 or 30lb by his tail, he was so slippery you'd drop it. So the only way you could do it was with a hessian sack. And if the water was above the putcher and there was a fish, you had to go down under the water, hanging onto the front. Thankfully in the few years I worked there I never encountered one under the water. But a chap called Peter Fresher used to work part-time for Norman like I did and he got caught on occasion to go down and check them and he slipped and lost one. But they come back in on the tide, they float round in the area. Of course, Norman went on the next tide and there was his fish. And he knew it had been dropped because it was white and it was floating around near the putchers. And he flipped.

Norman had one hell of a temper when things didn't go as planned. Comes from having a bash on the head after a motorbike accident. He'd just flip when things went wrong if we did things that weren't right to him. We did things his way. I remember coming back with the tractor and trailer, one time, loaded with more putchers from his farm when we were setting the putchers up at the beginning of the season. Coming back fully loaded, I drove the tractor down onto the beach and turned to come alongside the hedge – there's a hedge there before the corner.

He pointed to a photo of the putcher rank in Nick Large's book *The Glorious Uncertainty*, before continuing. The photo clearly shows how the putcher rank comes directly off the shore and curves round 90 degrees, pointing downstream, with a short hedge after the turn, running parallel to the shore, and then a hard corner with putchers out into the stream, perpendicular to the hedge and shore. Today, as I've said, the line of posts still stick out of water if you're there at low water.

I came round tight alongside, and the trailer was about 2ft wider than the tractor on either side and I caught the edge of the trailer on one of these spars that was sticking out and it knocked it down at an angle, one end on the floor. Anyway, we were working on the last batch and coming back up, Norman driving the tractor, and he saw it and he turned round on the tractor, which was still moving forward, and he started shouting and swearing and suddenly the tractor went down one of the runnels in the beach. It was stuck, wouldn't come out forwards or backwards, lying at an angle of 45 degrees in the runnel. And the tide was coming in round the tractor and I knew that if he lost that tractor he'd kill me. Really kill me. So I physically jumped off the trailer and lifted the front of the tractor, it was only a small Massey-Fergie, so I could move it about 2ft, and I was only 5ft and I did, I moved the tractor out of the groove in the rock. Otherwise he would have gone mad!

What about setting up then?

Well, we couldn't start before the 15th. We'd wedge them all in, wouldn't do it in a day because of the tide. Depends if it was neaps or springs but you'd get out as many as you could in the first day or two. The wind as well, as it stops the water going down. 600 he had. Brought them all down by tractor and trailer, all tucked inside one another. Several loads. If you did start say on the 14th, you had to place netting over the mouth to stop any fish getting in. We used long nails to secure them, all 600 of them, set into the gaps between the rails with the nose of each at the same level as the bottom of the mouth, angled down. They said those fishing the flood were angled more steeply as the ebb could wash the fish out. Then we'd get them out on 15th August. Check them as they came out. Norman would and some would be discarded. I reckon he'd make somewhere between 100 and 200 new ones each winter to replace those damaged or rotten ones. Obviously they rot being constantly covered and uncovered by the water. Though I don't think I ever saw him making new ones!

Taking them out obviously wasn't that easy, I thought. When I mentioned the bailiff, Steve remembered an encounter:

Did I tell you about the bailiff? I was there, I'd gone down on Norman's motorbike, which was a 250cc BSA maroon colour because everyone knew Norman's motorbike. I'd gone out and driven it along and parked it up on the rocks where the cliff is collapsing. I'd gone out on the putchers and I was actually stood on the rail, water swirling below me, and I'd almost finished, I must have been working my way back from the end, and a voice said, 'What do you think you're doing?' Just like that. And I turned round and I was eyeball to eyeball with him, and I'm on a rail. I thought, oh my God, who are you? I thought he was walking on water for a minute. Then I said Norman Knapp, that's his motorbike and I work for him. So I came off, when I jumped down onto the floor, he was 6ft 8in and he was the water bailiff. They'd had problems then with people stealing salmon. But you wouldn't want to argue with him. They used to get into terrific fights, those bailiffs, with people illegal fishing. And Norman has a short temper and was a big man, I wouldn't want to get caught messing around with his fishing. He had six children and the reason being that when he did actually make it home in the middle of the night, in the dark, his wife was so pleased to see him, she comforted him! She's still alive, Audrey.

The Knapps were a large family and Large's book mentioned some ten of them, excluding Norman. So, as a parting question, I asked whether he enjoyed it:

I was very fortunate, very lucky and I enjoyed it. OK, so you had to go into the water up to your chest, wearing waders. But the thing I did fear most was would there be a fish underwater if the tide didn't go out. But I was fortunate there. You forgot as it was only five months in the year. And it was only weekends. Sometimes, in the middle of summer when it was haymaking, I had to load all the bales onto the trailer seven high, on my own, with the tractor trundling along in gear with no one on it, going down the row, throwing these bales on, and he'd magnanimously say I'll go down and do the fishing for you. And I'd come back and find him watching tele. But I'd come back normally with two or three fish. They'd be brought on my back in a burlap sack, piece of baling twine around the bottom, slung around here and tied off the top, and the fish behind you. Sometimes coming back with a 100lb of fish on you. Yea. I very rarely came home without a fish.

Just along a bit, upstream on the same side of the river, but on the other side of the bridge, was the Folly Weir, the largest on the river with 800 traps and, at its outer limits it was six tiers of putchers high. Continuing upstream there were a host of putcher ranks between Littleton and Cowhill Pills, around the edge of the Oldbury Salmon Pool, and this where the story of three fishermen begins.

Every September, the Frampton Country Fair takes place at Frampton-upon-Severn and for the last twenty years or so Kipperland has had a stall there in the Living Working Countryside part of the show. And for many more years, Deryck Huby and Don Riddle have demonstrated the skill of making putchers to thousands of show visitors. These two fellows, now 95 and 91 respectively, both fished putcher ranks on the river.

Deryck moved to Gloucestershire in 1937 when he was 11 years old. His father had got a job as a gamekeeper on the Berkeley Estate and the family moved to an isolated cottage by the duck decoy pool on the estate. After school he became a farmer and went to Hartpury College to study agriculture. In the early 1960s, he joined the Milk Marketing Board as part of their insemination team and spent the rest of his working life in this role.

Don Riddle was born in 1930 and has lived in Oldbury-on-Severn all his life. He started fishing when he was 9 years old. The year before, his mother drowned in a tragic accident on the river and he went to live with his grandparents. After leaving school he also started farming until going to work at the Milk Marketing Board in 1962. Like Deryck, he also worked as an artificial inseminator, and he remained there for thirty-one years until retirement.

A row of eleven putts, nearly fixed, though without any hedge to lead the salmon into the mouths.

For most of that time, the two, who had become great friends, worked at the salmon fishing in the Severn. And from that time back in the 1960s, they've continued the partnership, not in fishing as such, but showing exactly how much skill and patience is needed to produce putchers, especially when you consider one rank could have upwards of 700 putchers in it.

Along with another friend, Merve Shipp, who also worked for the Milk Marketing Board, Deryck started fishing for mullet near Aust in 1963, and there they met Oldbury fisherman, Bob Knapp, who then encouraged them to take up putcher fishing by teaching them the craft of making putts and putchers and knitting nets. Don Riddle joined them as a threesome soon after. Deryck explains:

The old fisherman down there, old Bob Knapp, he taught us everything. Showed us everything. How to make the big kypes and that. Because he owned a rank of kypes, seventy I think it was in one rank. He had put them in once but it was on a flat bed. Of course, it was a lot of work. Yes, a lot of work. So we put in a dozen like that and only caught one salmon! You catch a lot of shrimps and eels and flatfish, which of course were saleable, but only one salmon. Then the following year we had a chance to fish some putchers. It was, um, they said, illegal putchers because there was no certificate of privilege for them. But what had happened was when they went to Gloucester, they had to go to Gloucester when they were issuing the certificates of privilege, that was in 1865, and this was on Thornbury Castle Estate and it was the last day of the hearings and the agent for the Estate never took the papers. They had the papers but he never took them to the hearing and it was too late, it was finished. But despite all that, in the Second World War, Bob Knapp and his brother-in-law used the lave-net there in the pool and there were also some putchers on the pool and he fished them too. He applied to the fisheries people saying there was a shortage of food and they agreed it could be fished under those circumstances. After the war they weren't fished but when we then applied they looked back and saw that they'd been fished and said, yes, we could fish them. It was legal to fish them. So we did.

By 1966, they were renting these three putcher fisheries between Littleton and Cowhill called the Lords Weir, the Lower Weir and the Gravel Weir. Together, these were 400 traps.

The putt, or kype as Deryck calls it, was a much more massive basket than the putcher and came in three detachable parts – a kype, which was the large basket at the mouth, a butt and a forewheel. It was detachable so that it could be transported down to the river more easily but was made so that the three parts would fit together snugly. It was constructed of local materials: hazel sticks and withies. Putts are the most ancient form of fixed fishing engine, with reliable records of them going back to 1533 and references to basket-type traps, which possibly were putts, back to 956:

We paid Thornbury Castle Estate so much rent and one salmon, and then so much for what we caught. It didn't work out too much. We had two fisheries to start with, Lords Weir where you could put 200 putchers in there but we only put in 150, and then there was Lower Weir and that was 100 putchers sat on the rock straight at the bottom of the Salmon Pool. Though when the tide got to 30ft at Bristol it blocked up with sand and seaweed for a week so we didn't get much from it. Also there was then Gravel Weir, which was further down and that belonged to Tony Bamfield at the farm. This was out across the Pool so that we couldn't walk across to it. Norman Knapp from Littleton had a canoe and he used to nip across. But you couldn't walk and we had to go out from Cowhill Pill and walk round. Anyway, Tony Bamfield had the Folly Weir and he said it was no use to me so you fish it and he wouldn't take any rent or anything for it. Just let us fish it and so we did that. It was an open one but we got a few fish in it, worth doing. We did that from 1966 to 1972. Put in stakes, though Gravel Weir had been fished before but a lot of it had gone.

In 1972, they then got the chance to move upstream to Hayward Rock, which was owned by the Berkeley Estate. Deryck's father was gamekeeper there for twenty-four years and I guess that helped. There they had the putts, putchers and lave-net fishing, one each. I asked about the lave-nets:

We had three lave-nets. The licences said the fishery started the 1st of February and finished the 31st of August. So what we did is I took full season licence out but the other two took a half season licence out and that started 1st May. I'll tell you about February. The best day's fishing was unbelievable. Originally all fishermen always used to put wire netting across where the putchers go. There were sixty putchers across the bottom in a V, and they'd replaced those putchers by putting some wire netting across. The old boy used to say he'd catch all my expenses before the season came in. So anyway, we asked the bailiff if we could put some hurdles across and he said yes. We made some up and put them in. It was the last day of February 1974, early one morning, a hell of a frost, and I just stood on the bank by the putchers, not been there long, and I looked down and there was like a little elver coming along. Grabbed the net and in and up. 31lb. And he laid in that net just like he was a log. Just as if it was frozen. Never kicked or anything so I killed him. Might be another I thought so I waited. Not long, ten minutes perhaps, bash in the middle, bang, water all threw up! So I waited and he came on round, I could see him coming round the head on the far side of the pool and next thing he's coming down towards me so he came and I scooped him up. 22½lb. That was the only fish we ever caught in February.

They fished the Hayward Rock fishery from 1972 to 1997, a time when they knew that the season was going to be cut right back from sixteen weeks to ten, and so they decided to finish. They were, as Deryck pointed out, the last to come into the salmon fishery on this side of the river and the last to leave it.

If you visit Hayward Rock today it's a very different place. The place where they used to park is a bit of a mess. The steps down from the embankment are still partially there but there's now a gate to stop anyone using them.

Deryck described how they walked down the first set, which are still there, and along a narrow concrete path, but the second set they used to access the riverbank were removed some years ago and replaced with a smooth concrete layer, which looks a death trap in the slippery mud. There's a deep hole there under the pool of water and he says it's best if everyone just stays away.

But I wanted to know more about setting up the ranks and making the traps. I remembered watching them work at Frampton and him saying a putcher would take an hour to make. However, just for the kype of the putt that was 6 or even 7ft in diameter that could take a day to make. And the work of setting them up is just as interesting. You can see the remains of the posts today at low water. He told me:

> Yes, 240 putchers there, two high. On the curve. And from the end of those, forward a little bit, was where the kypes [putts] were, a gap and then the kypes started. We could have fished twenty of them but we only put in about fourteen. And then immediately from the end of the kypes was this hedge, 300yd of hedge running out into the river, not quite straight. A lot of material. We drove in the posts with a drive hole, every yard and there we went over to Lower Woods in Inglestone Common and we had half an acre of nut [hazel] every year, which we then cut and hauled back with a tractor and trailer. Then we'd bundle up in about a dozen rods on the bank and it tie up and float it out when we wanted to put some in, leave it out there for a couple of days. Then we'd weave it in between the posts, not so dense but enough to stop the fish swimming through.
>
> You had to repair a section each year. Big logs come down the river and can break a stake. One year I remember, a cold one at Gloucester, about eighty-four I think, and we had a lot of snow. They tipped loads of the snow into the river at Gloucester and we had icebergs coming down. That did some damage, I tell you. Cor, it knocked some out then.

He paused while we pored over a map of the river. Pointing to Hayward Rock he continued:

> Then we had 60 putchers across the gap at the end of it, facing out, between there and the buoy. It used to be a buoy but now it's a light at the end of the rock. Hayward Rock. The salmon pool was in the middle and that never ran dry. We'd lave-net there where the water would be up to your chest.

Sometimes every day for one of us. Also at night when it was darkness and all that. But wouldn't go out if it was foggy, mind. And I wouldn't go out again if it was heavy thundering and lightning. I went out once when it came on, I was out there once with the lave-net and it came on suddenly. It was very scary. I thought I'd better get along the putchers so I was near the putchers and the lightning hit the rock not really far out. My God it was frightening. And I had three fish in the putchers all alive, and that always happens doesn't it. But I was glad to get out, I can tell you. The trouble is if it didn't hit you, you can lose your sense of direction easily, and the tide would be back before you knew it so you didn't know where you were going.

Of course people did drown in these circumstances and several lave-netters at that. For now, we will hear much more about lave-netting in a subsequent chapter.

Putts are fished through the season and then are blocked off with some spars to prevent fish entering, whereas all the putchers were taken ashore. Putts would last about three years. Taking putchers out, he said, was no problem as you could easily carry ten at one time, five on each shoulder. Taking them back ashore was different as they were much heavier. A laborious task. So I asked him about making the putts:

We'd put staves into the ground, six in a circle, then you go round so many times with withies to form the circle, then you put another one in each so that you have two, then when you come up a bit more you put three more uprights in, every fourth one, so that you'd have fifteen. What you do is do it on the ground and then you'd take it out and dig a hole so you've got it down there and you can work on the level.

Sounds easy! So we looked at photographs of putts and how they are assembled on the ground. Deryck pointed to what he called the forewheel apse to hold the forewheel in the fork. These are two small posts driven into the riverbed at an angle so that they form a fork. The same arrangement, though thicker and longer, held the butt, the middle section and then four upright posts or stakes, one either side of the mouth and another one each side near the end of the kype. The kype was tied to the posts using withies until polypropylene rope was widely available. However, for the forewheel and butt apses, a withy strap was tied across. In the case of the forewheel, this had to be easy to remove to empty the fish and shrimps. To stop the shrimps and eels coming out the bottom of the putt, they'd push a bunch of seaweed inside:

> We did away with that withy tie that had loops twisted either end to be looped over each fork and cut an inner tube which was much easier. All the putchers were tied on with withy ties, though we also used to use a piece of wood in the mouth fixed to the rail below.

The rail he referred to was one of the stout larch posts, some 40ft long, fixed to the stakes to carry the putchers. The longer the better as shorter ones wouldn't stay put for long. There was one rail about 9in off the riverbed at the front and another at the same level holding the thin end of the putcher, fixed to the elm stakes about a yard apart. Then another row of rails the same distance apart held the second tier of putchers. Nails, notches and ties were used in varying ways. They used to fetch the larch from Catgrove Woods on the Berkeley Estate, which wasn't too far away, and once nearly got in trouble with the police for using a tractor and trailer to carry them with some 20ft overhanging. Luckily he let them go with a warning: use a timber transporter next time!

'You never saw a police car down that lane ever,' exclaimed Deryck when I said I'd heard about that episode. Isn't that the thing about the police: you never see one when you need one and vice versa!

Deryck has an encyclopedic memory about their work at the salmon and talked passionately about it. It is hardly surprising, as he and Don spent thirty-four years at it together, Don for longer. The fact that they, in their nineties, continue to exhibit at Frampton each year as a reminder to folk of this lost way of life is surely a measure of how much they immersed themselves in it.

Don wasn't able to join us at Deryck's house so I thought I had better call him to see what he had to say. But his hearing aid was causing him problems and he couldn't understand what I was saying. Then the coronavirus lockdown hit hard in 2020 and all contact had to stop, to which was added the fact that Don's hearing aid broke. Phone communication broke down too! On top of that, Frampton was cancelled. However, my mind cast back to that show and watching Don and Deryck at work, each explaining their task.

I can picture Don splitting a willow rod into three pieces using his oak cleaver, which, he said, was his father's. He and Deryck work off a home-made bench, with the cleft willow rods stuck into the nine pre-drilled holes around a circle of a diameter of 10in or so that slant inwards so the willow bend outwards, creating the funnel shape. The cut willow is pushed through each hole into a much smaller diameter withy ring sitting on the ground below the bench, 18in lower. Then a ring is plaited at bench level and shorter pieces of cleft willow are then placed between the nine, wedged into the ring. Another ring is then plaited about 15in higher and another at the mouth of the putcher, thus creating the entrance at about 24in in diameter. The putcher can then be removed from the bench and turned the other way up, and the lower part plaited in a spiral fashion to end up at the tip of the putcher, where another 4in diameter withy ring completes the putcher. They used to say a putcher took an hour to make and lasted two years, maybe three at a push. The ones they make today are more likely to go to some-one's garden as an ornament than anywhere near the river, though they do have some in various places such as Slimbridge Wetlands Trust as exhibits. Though I guess when these two stop that will be yet another countryside tradition gone out of the door, just as the fishing itself has.

Talking of the end of fishing, local fisherman Nigel Mott, being one of the last surviving proponents of putcher fishing, had a bank of metal putchers near Lydney, from which it was reported he made £60,000 annually with his mate, David Merrett. Then along came the EA in 2012, who then ordered him to cut his catch from 600 to just thirty fish a year. Mott believed that by slash-ing his permitted catch this made his putchers wholly uneconomic to work, thus resulting in in his lease becoming worthless. So Mott took the agency to court on the basis that this decision violated his human right to peacefully enjoy his private property. In 2015, a judge found in Mott's favour and the same judge, Judge David Cooke, in the Supreme Court, ordered that the EA pay him £187,278 compensation for the destruction of his trade. The case went to the Court of Appeal, where judges found that, although the EA had grounds to reduce the catch limits on environmental grounds, they couldn't simply take away his livelihood without compensation. More damning was that they found that there was absolutely no evidence that the agency had even considered the devastating impact on Mott and his livelihood. As we shall see over the coming pages, the government quangos – the EA, NRW and Marine Scotland – simply act in undemocratic dictatorial ways, without any suggestion of negotiation and with intentions that are at best opaque and certainly not what they purport to be. More like protectionism!

8

LONG-NETTING THE RIVER SEVERN

Long-netting was another form of salmon fishing on the river and was, in general terms, the sweeping of the river in the hope of netting a salmon swimming upriver. In other words, a form of the seine-net. In this instance on the Severn, a punt was needed, these being flat-bottomed, square-shaped boats of a very ancient design, and built from pitch pine, oak and archangel whites. These able craft were capable of working in very shallow water and could also carry very heavy loads. Usually their work was not confined to fishing as they were just as likely to act as ferry, animal transporter, reed-cutting boat and general work craft. I say was, for long-netting is no longer but luckily my mate Simon Cooper has the last one of these known to exist in a floating state, and way back in about 2002, I think, we went fishing just the once. Indeed, it was Simon who had instigated the whole thing on that memorable, fine summer's day.

Long-nets, just like the seine-nets of the Appledore rivers, were set at particular stations out from the riverbank, there once being some twenty-six set positions on the river between Tewkesbury and Framilode. Below Framilode, it was only the domain of the stop-nets and lave-nets due to the nature of the river. The long-nets themselves, more costly than stop-nets, were up to 200yd long and were made to suit the river at the particular station as they shouldn't be longer than three-quarters of the width of the river at that point. However, in their desire to conserve stocks in 1952, the Severn River Board closed all the long-net stations above Gloucester. Even with the remaining stations below the city, this was probably the least popular way of fishing due to the lack of fish. It is said that it was often difficult to find four men with the enthusiasm to bother with the hard work. Enthusiastic as we must have been, we were lucky then!

Shooting the net from the punt.

We were using Simon's punt, into which the net had already been placed on my arrival. I was late after getting lost and missing the lane down to the house belonging to Frank at Bollow on the west side of the river, above the Arlingham bends. Alan, who held the licence, was to be in the punt along with Simon and Gordon, while Dexter was acting as the debut-man who would be the one to walk along the riverbank with one end of the net during the fishing process.

Hauling in onto the far bank.

So, when the salmon were judged to be swimming upriver against the ebb, fishing started. The punt was rowed out to the far bank to drop Dexter off and for him to take that end of the net, this being paid out over the stern as they went. Actually, I say rowing, though Simon had fitted a tiny electric motor outboard that helped in this matter. But it wasn't used once the punt was in the deeper water for fear of frightening a passing salmon.

Dexter stood still on the water's edge on the east bank, holding onto his end. Once the whole net was set round in a clean circular arc, which is vital to prevent the tide ruining the run, that end of the net, the muntle, is held in the punt while the whole lot drifts with the current, Dexter walking at the same speed with a canvas harness around his chest and shoulders, thereby pulling his end along. It was vital he kept the end of the net close to the riverbank to prevent a salmon trying, as is their normal behaviour, to swim around the net rather than heading back downstream. This is also the main skilful time, ensuring the arc stays a clean sweep and recognising if a fish is in the net. It's also important not to net the entire width of the river if it narrows, which is against the regulations. By this time, Gordon had joined Dexter on the far bank and when the run of the net was judged as being to the extent of the draft (that part of the river they are allowed to fish), the punt was rowed towards the shore and a line weighted with the dud-jack was thrown ashore to Gordon, who was acting as muntle-man and was down-stream of the debut-man at this point. The net was then closed and the punt beached for Alan and Simon to join in with the steady haul of the net.

At this point, I was able to join in with the pulling of one end of the net. Dexter held the bottom edge of it to the riverbed to prevent any fish escaping that way. The excitement grew as the net reduced in size, especially as Alan thought he detected movement. It could easily have been a mullet as the river was full of freshwater, not great for salmon. Suddenly, there was move-ment in front of our eyes and the top edge of a salmon appeared. He was swimming upstream as the net was hauled and Alan swore as he had forgot-ten the ring-net to land the fish. He splashed into the net, water spraying as he tried to flick the salmon onto the mud. Luckily he managed to just before the fish swam over the top of the net, and he lay flapping on the mud whilst Alan found an old brush with which to finish him off, the priest also being forgotten. Should we shoot again? Yes, but the second time the net was empty. Still, with a 12lb 9oz salmon, Alan was happily soon on his way up to the Severn & Wye smokery to sell his catch.

There are variations in the method, so I'm told. Sometimes the debut-man stays put while the punt rows out at an angle across the river and the muntle is then passed over to the muntle-man, who then walks it downstream on the opposite bank before returning to the punt and crossing back across to then draw both ends in. I guess it all depends on the strength of the current and the vital thing is to ensure the net stays 'sweet' and avoids any obstructions, with the leaded footrope running smoothly along the riverbed. The lie of the corks tells the fisherman all he needs to know.

However you look at it, by the 1980s long-netting was only practised by a small band of dedicated people largely because, although no longer eco-nomical, they, usually elderly folk, had always done it. As Nick Large says, 'It persisted because of a sense of tradition,' as we shall see with many of the ancient forms of fishing this west coast has.

At Newnham and Awre it is believed that boats similar to the stop-net boats, of which we will learn more in the next chapter, worked long-nets. One flat-bottomed, heavy boat used at Framilode, similar to a stop-net boat, was built at Saul Junction on the Gloucester and Sharpness Canal for the Gower family. However, Scottish cobles, between 18 and 20ft long, were also brought in about 1910, and several 17ft cobles were still in use at that time. These were said to have been introduced from the east coast of Scotland about 1900 to the River Wye by a fisherman called Miller, who then moved to the Severn ten years later.

North of Tewkesbury, coracles were once used to take salmon, with Ironbridge, Shrewsbury and Welshpool all having their own peculiar type of coracle. The Salmon & Freshwater Fisheries Act of 1923 ended this coracle fishing because of the lack of salmon in the river. Today's stocks are said to be equally as low, which accounts for the fact, as elsewhere in England, that there's no commercial salmon fishing in the English part of the River Severn any longer, beyond the occasional poacher! But then again, as we've seen so far, and will learn much more about, who can actually trust the words of the EA when they've so many vested interests to take into account, some being very close to their home ground? We will hear about their mantra of 'enhancing salmon stocks' but with very little action to show. For now, salmon fishing on the upper reaches of the river is a thing of the past.

9

THE STOP-NETS
OF GATCOMBE

In about 1998, I was lucky enough to visit Ann and Raymond Bayliss at Court Farm, Gatcombe, largely thanks again to my friends Simon and Ann Cooper, who readers will remember from the previous chapter. Hidden away from inquisitive travellers and surrounded by a shroud of trees, this tiny vale edges the lower part of the River Severn, a few miles above Lydney. It remains as quiet and unspoilt as it has for many generations, generations that have seen an active participation in the salmon fishery.

Between the half dozen or so houses and the river, built upon an embankment, is the main railway line between Gloucester and South Wales. Below this is an archway that enabled boats to be drawn up out of the river and safely kept ashore. And just alongside this, pulled a bit up the bank, I came across three stop-net boats, which, although they'd seen better days, were the last of a type that was in constant use in the river during the salmon season up to about the mid-1980s.

Stop-netting was a particular method of catching salmon in the lower reaches of the Severn. Above this point the river gets narrower – north of Awre – and long-nets were used to take salmon, and coracles, even further upwards to its source, but these weren't suitable on the more tidal section. So evolved the use of a stop-net.

To stop-net one needs a boat – a stopping boat, like those on the shore. These are 20–22ft long, 8ft in the beam, open boats, very sturdily built, heavy to manhandle on land yet easy to scull with one oar when on the water.

After the dissolution of the monasteries, the fishing rights came to the Lord Bledisloe of Lydney (Wellhouse Bay) and Crown & Hygrove (Gatcombe), and in 1794 the tenant to these rights was George Smith. He was followed by John

Shaw, who passed it to his son William. In 1878, William Shaw, his brother Thomas and a local fisherman, Thomas Margrates, were bringing their stopping boat upriver when it hit the nearly completed Severn railway bridge that, when opened the following year, connected Sharpness and Purton. Unfortunately, William Shaw was drowned, so that Thomas, to whom the fishing rights passed, decided to sell on the lease. These were bought by Charles Morse that same year and remained in the ownership of his family, passing eventually to his great-granddaughter, Ann, in the 1970s, and her husband Raymond Bayliss, the last fisherman who gave up the method a few years before our visit.

Fishing in the river wasn't the subject of much control, except for paying the rent, until the Special Commission for the Salmon Fisheries reported in 1866. William Shaw gave evidence to the panel and established his rights to continue fishing at seven chains in nearby Wellhouse Bay, as did the other established fishermen on this and other rivers. In all, twenty-four nets were authorised in the river for use by him and others. In Wellhouse Bay, these were each named as Hayward Rock, Long Ledge (where there were two), Round Rock, Fish House, Old Dunns and The Flood. Each chain had a wire cable stretched out between an anchor into the deep water and a stake on the beach at low water. The boats would be attached to the cables, with up to six boats moored on any particular cable at any one time. Ten stopping boats were based at Gatcombe. Up to 1937, as we've already seen, the family also had rights to fish in the river at the Hygrove fishery at Gatcombe itself. However, this part of the river was more dangerous than Wellhouse Bay, and stop-nets ceased to be used hereabouts when one fisherman, F. Fenner, was drowned while fishing.

A stop-net was basically a net suspended from two rimes from a stopping boat, which was moored across the river upon the chains, beam-on to the current and wind that sometimes opposed each other to create dangerous seas. It was a bigger version of the pitch-net we saw on the River Parrett, if you like. These rimes were big and heavy, some 24ft long and made from Norwegian spruce, and were held apart by a spreader so that the mouth of the bag-net was some 30ft. The frame was lowered into the water, into the tide, so that the net opened up underneath the stopping boat, a forked prop positioned to support the rimes. The net end was on the downstream side of the boat and from here, five feeling strings were attached to different parts of the net and to a 'tuning fork' – a wooden stick – held by the fisherman.

The net is raised onto the gunwale of the stopping boat and a salmon removed.

Weights were attached to the apex of the rimes, so that once a fish was felt in the net, the prop was kicked out – called 'knocking out' – and the mouth of the net, helped by the weight of the fisherman, brought to the surface. The end of the net was quickly inspected for a fish, and if one had been caught it was extracted by undoing a cord that opened the 'cunning hole'. The fish was killed by a quick thump on the head with the 'knobbling pin', and the cord retied, the net passed back over into the water, and the rimes lifted and propped up to lower the net once more.

The fishermen have to remain silent for the period they fish, probably about three hours at a time. Like any type of fishing, sometimes they might go for days without a fish, especially when the tides are strong or the sea flat calm. At other times, in favourable conditions, they might get two or three or four. The best condition was a mixture of a good southwesterly wind and a 21–22ft tide in the river. In Nick Large's book, he tells of the time he went out on two occasions, firstly in 1969 with Christopher Morse and again in 1975. On the second occasion three boats went out fishing and they caught fourteen salmon!

By the year we were there, stop-netting hadn't been practised for more than a decade, though the signs that this small hamlet had once thrived upon the fishing were still dotted around. We had called in at Court House and met Ann and Raymond. Ann showed me the fish records of both her grandfather and father, leather-bound volumes that gave details of differing aspects of the business. The 'salmon accounts' detailed each fish caught, its weight and price fetched, while the wages book gave the earnings of the stop-net fishermen responsible for landing these fish. Lave-nets were also fished here, of which we will learn more in the next chapter, and another volume named each lave-net fisherman and the number of salmon they took, while his accounts book listed all his own annual expenses. Each entry was handwritten in a flowing style, and each gave a special insight into the mechanics of the fishing. The boats were owned by the Morse family, who also paid the rent and licence fee. In return, each fisherman gave them every salmon, which they then sold. The fishermen got about half the proceeds from the sale.

In the outside workshop, I remember Ann showing me where she used to help her father, Christopher Morse, make up the nets. Before 1965 these were hemp nets, the raw material being purchased from Bridport, and were approximately 76ft long by 30ft deep. The top 3ft of the net had a 7in mesh, decreasing every 3ft to 6in, 5in and 4½in and then the bottom 18ft having a 4in mesh. Ann herself recalled assembling the shaped net by knitting various meshes together. The different-sized mesh pins, wooden needles and the wooden winder were still there, among the boatbuilding tools. On the lid of the net chest, two dog-eared faded pieces of cardboard still hung, dated 1899 and 1923, which listed the name of each fisherman, the particular boat they used and the size of net they required. It was like nothing had been touched for over a century, a real-life museum!

Outside the blacksmith's shop was the old copper boiler, into which the net was put for tarring. Here the fishermen would immerse their nets into the mixture of tar and tar oil for a few days, before hauling them out using the pulley that was still suspended above, and leaving them to drip for a further three days. Then they were hung up to dry thoroughly. However, nets rarely lasted for a season, so they were only 'barked' at the beginning of the season, and not during. In fact, the whole place reeked of 'museum' and some twenty years later I still wonder what happened to all and sundry.

Christopher Morse had taken over the running of the family business in 1931 after Charles' death. He was 20 years old at that time, and he spent a lifetime at Gatcombe after being born in Court House, where he remained right up to his death in 1972. After that, Ann and Raymond took control.

Once legislation was introduced to exercise some control over salmon fishing after the 1866 enquiry, fishing for salmon used to begin traditionally on 2 February, when the boats were first launched. Fishing would from thence continue right through the season until its closure on 15 August. By the time stop-netting had ceased in 1984, the season was down to a short summer of fishing.

'Bringing' the boats in at the end of the season was a tiresome task for which horses were used until tractors appeared. On the August high tide, just after the full moon, all ten of the boats were brought in under the railway line, one by one, and hauled up the slope to be blocked up for their winter stay. Once they had dried out thoroughly, they were repaired, tarred and painted ready for the next season. Ann Bayliss remembered this as a time when all the fishermen and their children joined in with the work of hauling out, a time of celebration after a successful salmon season. A celebration of the annual harvest; a form of thanksgiving for both the salmon and the fishermen's lives. After the work was completed, all the helpers went up to Court House to take of a meal of cold salt beef, beetroot and mashed potato, followed by home-made apple pie, accompanied by the fishermen's stories that rolled on and on ...

She also remembers her father telling her of the building of the last stopping boat in 1922. This was the *Margaret*, No. 30 – they were all numbered – which was built by her grandfather and fisherman Joe Wathen. This was named after her father's youngest sister. Today she remains as one of the three boats hauled up the slope alongside the last house of the hamlet, almost rotted away.

They bought most of the oak for the keel and ribs at nearby sawmills on the edge of the Forest of Dean. Each piece was then cut to shape by hand in the workshop, outside of which the growing boat was sitting. Ash was used for the gunwale, and elm for the planking below the waterline. Above this, larch was used for the topsides, thwarts and short decking in the bow. The finished boat was then caulked, pitched and fitted out with ironwork that they forged themselves. It took something like three months to build, although exact times are unclear.

The same sawmills were visited if a plank was needed to be replaced in another boat. Sometimes this task would take days as there was no electricity at Gatcombe back then. Each plank was steamed in the steam box that was heated by the forge. Ann remembered one occasion when her father spent days making up one plank in the carpenter's shop, which then split on the last nail. Everybody made themselves scarce after that for a while!

Raymond Bayliss in his boat, showing off a salmon. See how the boats were stretched out in a line on the chain. His net is raised and the weight fixed to the rimes can be seen on the bottom boards while the nets are lowered on the other two boats.

Winter time was as busy a period as the fishing season. As well as repairing the boats and the gear, the Morses made a living from general carpentry repairs and making ladders to sell. They also went plum-picking and made cider, of course, this being Gloucestershire! They had cows, too, that needed milking and calves to be tended by hand. Ann recalled milking fourteen cows and taking the churns up to the top of the lane to be collected by Cadbury's before going to school. While I was there that time Raymond showed us a side of his home-cured bacon and the apple press used for cider-making still sat in the same shed. We bought a bit of the bacon off him and it really was delicious when later grilled over a camp fire.

Ann's grandfather on her mother's side looked after the lights on the Severn railway bridge, while Charles Morse ran the local Blakeney Gas, Light & Coke Company. Another winter job, usually just prior to the season commencing, was the laying of new galvanised cables in the river at the fishing positions. I can still picture the note hanging in the carpenter's shed, dated 1915, noting the length of these in fathoms: Gatcombe – 83; Fish House Head – 73; Low Water Wire – 45; Middle – 83; Dunns – 65.

The fish house referred to was at Wellhouse Bay and was a stone building with two rooms – one with bunks, cupboards and an open fire and the other that was used to store ropes, anchors, weights and other gear. The fishermen would sit or sleep here, waiting for the right state of tide to begin fishing. You can imagine the stories and yarns they'd tell each other while the wind whistled and the Severn Bore charged past. When the fish house was destroyed in a particularly high tide about 1950, Christopher Morse built a wooden shed higher up the railway embankment. Although parts of the remains of the stone walls of the original house still exist, all signs that there ever was a wooden shed have long gone.

The Morses bought all of the fish caught in their stretch of river from the lave-net fishermen, as well as handling the stop-net salmon. In the nineteenth century, the fish were sent to Stroud and Cheltenham, and later to Billingsgate. For the latter they were packed in wooden boxes and sent directly by train from the station at Severn Bridge, then Awre when that closed, then Lydney until they refused to take goods, and thereafter Gloucester. For individual fish, they were sewn into bags called frails with fern from the woods to keep them damp. At one time, they were sent in willow baskets made for the purpose. According to a list of fish buyers dated 1915, they were sending it by train to Birmingham, Wolverhampton, Walsall, Warwick, Stourbridge and Leamington Spa.

Taking into account the amount of money paid out to the fishermen, and the upkeep of the boats and gear, it's hard today to see how the Morses made a living from their fish business. Looking through the record books gives a fantastic insight into the workings of it, and we can ascertain most of his expenditure and the price obtained at market for the fish. For instance, in 1913, the average amount received for his fish per lb ranged from 2s 3d in February to 1s 6d in March to about 1s 2d–1s 3d in July. The average wage for the fishermen was about 8s per 10lb caught. The salmon themselves averaged in weight from about 12–20lb. The total number of fish caught month by month in 1913 was: 4–24 Feb, eleven fish; 4–31 March, eighty-seven; 1–30 April, seventy-eight; 1–31 May, 206; 2–30 June, 468; 1–31 July, 272; and 1–14 August, fifty-seven, giving a total of 1,179 fish. Total fish sales for that year amounted to £1,070 13s 5d, with allowable expenses (boat and gear costs, wages, transport, rent and licences) of £646 14s 3d, giving a profit of £423 19s 2d. And that was a good year! A licence that cost £3 in 1893 had risen to £12 by 1961.

But that's all gone. And since our visit in 1998, so have both Ann and Raymond, who no doubt are looking down upon all the rod fishers and wondering how come those that fish as a sport are still allowed to fish when those that made a meagre living from the river have been confined to the past. So totally, totally unfair and immoral.

Salmon netting in Sprat Pool, which lies in the middle of the Taw/Torridge Estuary close to the confluence of the two rivers, using a hemp net, c. 1960. (North Devon Maritime Museum)

Stephen Perham of Clovelly hauling in the top of his drift-net with the floats, with Peter Braund bringing in the leaded foot (bottom) of the net; a shot that resulted in a good haul of herring. (Charlie Perham)

Stephen Perham and the author picking herring from the nets on Clovelly beach, 2018. (Ana Smylie)

Adrian Sellick emptying his nets out on the sands of Bridgwater Bay.

Shrimping with the Weston Bay Shrimpers, autumn 2019.

Deryck Huby and Don Riddle (bent over) building up the hedge to lead the salmon into the traps. Putts are in the distance. (Deryck Huby)

Deryck Huby holding a salmon from one of his putchers. (Deryck Huby)

The start of the haul of the long-net in the Severn, Alan Osment on the right of the photo, Gordon (middle) and Dexter (left). This was about 2002.

The lave-net fisherman's sculpture at Black Rock, unveiled 2020, a fitting memorial that, at publication, might be celebrating the end of this heritage fishery.

A postcard of two coracle fishers, showing off their net, in the 1950s.

Alan Lewis preparing to shoot his compass net into the western Cleddau, in 2015.

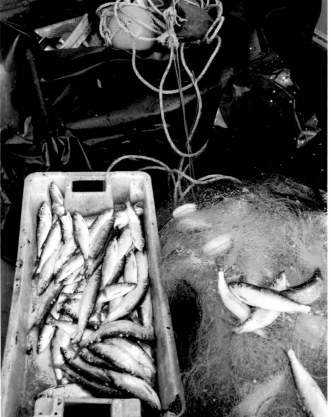

Fresh herring straight out from the River Cleddau in February 2021. (Alun Lewis)

Salmon seining in Portmadog harbour in about 2002. Unfortunately, there were no salmon, or indeed any fish, caught this day!

Standing on the stern sheets of a Dee salmon boat bringing the net aboard from the shore.

'The Last Shrimper', the sculpture in Lytham St Anne's Lowther Gardens, said to represent Russell Wignall, who is indeed the last full-time shrimper.

Salmon boats moored on the south bank of the River Ribble with a shot – identified by the cork floats – in the process of being taken back to the shore to where a man awaits.

Margaret Owen, in her 'yallers', on her way down to the river at Sunderland Point with her haaf-net.

Tom Smith hauling in his whammel-net out of the river in the 1950s. The lighthouse at Plover Point is on the left of the photo. (Alan Smith)

Michael Wilson showing me his lave-net that he fished with for many years. It's a smaller version of the River Severn lave-net for use in shallow water. His father's net was in the shed too.

A salmon in the haaf-net of the Annan fishers. (John Warwick)

A view inside the Snaab stake-net when it was still in use. (John Warwick)

A sunset view of the Upper Kilbride Fishery stake-net after Ana and I had trudged through the mud, fearful of the rising tide.

A terrific nostalgic view of Mallaig and its array of wooden fish boxes on 23 July 1969. (Linda Gowan)

A view of the cottages at Camas, Isle of Mull, with Allan and Maimie. (Sandy Brunton)

The catch being brought ashore at Camas. (Sandy Brunton)

The coble Iolaire, OB226, as we found her by chance at Kilchoan in the summer of 2019.

The bothy at Cuil Bay in the 1950s. (Sandy MacLachlan)

Mending the nets on the beach at Badentarbet, just north of the hamlet of Achiltibuie, where Peter Muir's father and uncle shared the bag-net fishing.

The kids and I at Red Point, on the north-west entrance to Loch Torridon, in 2019 exploring the old rusting anchors and the old buildings of what was once a salmon station.

Looking down towards the landing slipway at the Armadale salmon station, c. 2008.

The ice house and other buildings of the Bettyhill salmon station on the eastern bank of the River Naver, c. 2008.

Hauling up a bag-net to dry on the net poles on the grass by the shore near Castletown, between Thurso and John O'Groats in c. 2008.

10

FISHING THE RIVER WYE

As the fifth longest river in the country, the River Wye was once considered among the finest salmon rivers. It runs from the 2,000ft-high slopes of Plynlimon in Mid Wales, flowing some 150 miles down to the Severn just below Chepstow and is certainly the best in Wales. At its peak, rod catches of salmon exceeded 7,000 fish a year. Robert Pashley, who fished a stretch near Goodrich Court, some 4 miles downstream from Ross-on-Wye, was the most successful salmon fisherman in the history of the Wye and he caught more than 10,000 salmon in a forty-five-year period between 1906 and 1951, his best year being in 1936 when he caught 678! All on a rod!

But, as always, legislation controlled the fishing and arguments between commercial fishermen and the anglers persisted. After the passing of the 1861 Salmon Fisheries Act, the following year the Wye Preservation Society was formed to enforce the laws from this Act. At the first meeting in December 1862, Sir Velters Cornwall of Moccas was elected Chairman and President of the Society. This society, four years later, was then replaced by the Wye Board of Conservators, who were empowered to levy licences. But some of the keener rod fishers were still unhappy with the fact that the commercial fishermen were still able to net salmon and thus the Wye Fisheries Association was formed in 1875, with the sole purpose of obtaining control of net fishing, which predominated on the river. In 1878, John Hotchkiss joined the Wye Conservators and he continued to battle against commercial fishing and eventually was able to buy up most of the netting rights that, due to a dearth of salmon, had almost become worthless.

In 1901, the Duke of Beaufort sold his netting rights to the Crown for a generous amount of £5,000, and these fisheries were then taken over by the Wye Fisheries Association, who then let them out to anglers. By 1909, all netting above Brockweir Bridge was prohibited.

Drifting for salmon off the entrance of the River Wye.

The Wye River Board replaced the Wye Conservators in 1952 and it was they who then controlled the fishing in the lower reaches of the river, issuing thirty-five stop-net licences and, it appears, three putcher licences.

These commercial salmon fishermen fished either drift-nets or stop-nets and these were mostly operated around Chepstow under the watchful eye of the Wye River Board. Mike Taylor comes from a family of stop-net fishermen who lived at Beachley, which is the spit of land across the river sandwiched between the Wye and the Severn. Mike started his fishing career when his cousin Jack was taken ill and he was given the chance to take over Jack's drift-net licence by the then bailiff, Owen Hudson. He says it was a great adventure to fish out in the estuary in all weathers. Generally drifting started in February, although sometimes it was too cold. It occurred on neap tides when the river wasn't running fast. Their nets had a 6in mesh to catch the spring fish, although during the mid-summer they changed to a smaller mesh for the grilse and he remembers catching maybe twenty fish on a good day.

They would start fishing beneath the Severn Bridge, covering the mouth of the river so that when they drifted out of the river the net would be at right-angles to the shore and then they'd drift as far as Black Rock. Go further and they could get forced into the riverbank as the motors wouldn't be able to fight the current until the flood. They'd then get another drift on the way back upstream. You could tell when a salmon hit the net by the way the floats bobbed about in the water and you could sometimes feel the vibrations. Hauling in could be problematic because of the rudder.

Just as on the Severn, stop-nets were used on the Wye by the commercial fishermen and they worked best at spring tides. The origins of stop-netting seem obscured in history, though, according to myth, Wintour's Leap was named after Royalist Sir John Wintour who, hotly pursued on his horse by Cromwell's forces, jumped off into the Wye and swam to safety. However, what happened in 'reality' is somewhat different as, although he did escape from the Roundheads, it was in a less spectacular way by escaping down a hidden path and hailing a salmon fisherman in his stop-netting boat. This would have been in early 1645, which suggests stop-nets are much older than previously thought. Myth or reality: you can decide!

Like the Severn boats, they were heavily built in the same materials (oak, larch and ash), though slightly smaller at around 20ft in length and 7ft in beam. Unlike the Severn boats that moored along lines, they were generally fastened to the fishing stations using 30ft-long poles pushed into the riverbed, holding the boat against the strong current. Some fished as far upstream as Brockweir. Fishing was between April and August. Much of the fishing occurred off Chepstow, where some eight boats worked up to the end of this mode of fishing in 2000. To fish, the boats moored in the middle of the river – half in England, half in Wales, so to speak, held against the current by the poles stuck into the mud. A fine description of the way the boats were set was given some years ago by Geoff Bollen to the Chepstow Museum, who recorded him, and who began his account on where he would find his boat moored at the bottom of the steps before the Chepstow Fisheries. It's really worth visiting the museum just to hear Geoff's oral evidence of how the fishery operated.

Geoff had a rock that was his marker to drop his anchor, lining up the boat exactly before positioning the boat across the current with three poles, the first one pushed down over the gunwale and lashed. The second pole went in over the stern at an angle of about 45 degrees and was lashed. The third one was again lashed to the stern and then the first one moved to the bow of the boat, tilted in at a steeper angle and tied off. Thus all three poles were holding the boat across the current of the river before the anchor rope was released gently to ensure the poles were holding the boat. His warning was not to do that too fast, otherwise the poles might kick out and it would be back to square one.

The next task was to prepare the rames by spreading them out into a 'V' shape, some 25ft at the outer ends, the tension created by pegging the spreader. Then they could be lowered into the water, slowly as this adds weight to the poles, with the net flowing out beneath the boat. The rames were held up by a prop from the bottom of the boat. Then you could start fishing, using the strings from the net to 'feel' for a fish. When there was a sense that there was a fish in the net, the prop was kicked out, the rames were pressed down and the net lifted out of the water. To get any fish out, the net was gathered up and the fish killed with the priest before dropping the net back into the water and setting the net back underwent the keel.

It was, in all, quite a telling process, dangerous and skilful. On setting up a sheer in the river, as it was called, Mike Taylor commented on what he regarded as being a complex process. Once the boat was anchored, he said, if you didn't get the first pole in right, then you'd have to haul it out. These poles were 30ft long, weighted at one end, and often you had to pull the boat out and set it up again by tying with a rolling hitch to the stern.

In the river the fishermen had names for the fixed berths: Gut, Gloucester Hole, Slip and Cliff, Port Walls, Behind the Plane, Inside Castle, Outside Castle. Some are fairly obvious but 'Behind the Plane' remains a mystery to me!

Stop-netting seems to have been more successful during drought conditions, when the fish tend to stop back due to a lack of oxygen in the water. During a spate, with more fresh water, the fish don't drop back but swim upriver. The argument for stop-netting was that if they didn't catch the fish then they'd probably die anyway due to being in poor condition as many dying to dead fish were being seen in the river during such times when the river was low.

In 1995, the Wye and Usk Foundation was formed as a registered charity, with the remit to address the concerns of the decline in numbers of Atlantic salmon in the rivers. This was the largest river trust in the country. Five years later they bought out all the remaining commercial fishing licences in the river, in 2000 bringing an abrupt end to stop-netting and its 350 or more years of tradition and resulting in both Geoff Bollen and Mike Taylor, and the other stop-netters, stopping fishing altogether as drift-netting had already ceased by that time.

Today the frontage of the river is lined with a hopscotch mixture of plastic boats. The Wye Fishery Board building remains, though appeared empty in November 2019. The 'Hole in the Cliff', across the river, belies its position.

The quay still juts out into the river opposite the Boat Inn, where Geoff and the others used to keep the stopping boats. Fixed to the wall of the river is a plaque entitled 'The Decline of the Port' with some information about shipping and the fishing, just one in a series of sculptures and notices reflecting the importance of salmon fishing in the town. In reality, this is the only sign that there ever was any form of fishing here. Around the town there are other signs: lovely carvings in the wall by the cenotaph and a plaque outside the Tesco superstore. But my favourite has to be a sculpture by Barclays Bank with the perfectly apt words: 'A Chepstow salmon is worth its weight in gold, unlike the slab fish in London sold.' Says it all really! Take a walk and use your eyes as there are other little subtle references to just how important the stop-net fishing was to the town. All bravo to the salmon!

Today, the river continues to flow in its milky brown colour and will always wash against the mud-strewn bank. A neat bit of grass has 'Private' signs, which inform that this bit of the riverbank belongs to residents from a nearby block. Further along to the Old Town Bridge, the Riverside Wine Bar suggests an up-and-coming residential area. Somebody, though, has remembered that people still venture onto the river as along towards the A48 bridge, there's a sign letting us know that 'Rescue Equipment' is still available.

Upstream, rod fishing continues to thrive, albeit with lower catches. Until recently, only the lave-netters of Black Rock continued to fish in a commercial way anywhere near here, with licences issued by NRW. They recently declared that more than 60 per cent of protected rivers in Wales exceed phosphate pollution limits, including the Cleddau, Eden, Gwyrfai, Teifi, Tywi, Glaslyn, Dee, Usk and Wye. Along with salmon, these rivers support some of Wales' most special wildlife including freshwater pearl mussel, white-clawed crayfish and floating water-plantain. Currently, minimal regulations control farm pollution through voluntary 'good practice' guidelines, which aren't working.

So, for the time being, the Wye commercial fishing has all gone. All that we are left with are the professional statisticians, who insist that salmon stocks are still on the decline. Amateur video evidence seems to contradict this but that's no great surprise when the riparian owners are still selling riverside fishing at inflated daily rates! Confused? I am!

11

THE LAVE-NETS
AT BLACK ROCK

Black Rock sounds slightly foreboding in terms of sea talk and conjures up images of reefs lurking beneath the ocean. Dubh Sgeir in terms of navigation. But in terms of river talk, I was unsure what to expect until arriving at the bleak beach, one wet afternoon in early summer, a few years ago. The first thing I discovered, after parking up in the convenient car park, was the restored fish house that the Black Rock Lave Net Heritage Fishery had made their home in. I was somewhat early but this was already open and I found Martin Morgan, the Association's fish secretary, inside, getting his gear ready for fishing. A big, strong fellow with that South Walian soft lilt, he was also keen to talk fish, and so the minutes flew by until other members of the team arrived to interrupt the flow.

The second thing was that, every time they caught a salmon, they hung a wooden one up in the tree by the shed. Four hung, silent and still. It was late in the season, so was this an omen? Would we be lucky or was it a matter of Nick Large's glorious uncertainty?

The tide was high enough for us to walk on to the muddy, not-so-black foreshore, to the small fibreglass dinghy they moor up and which had just floated. In we went, five hefty lave-netters and me. There would normally be six but one had kindly foregone his chance to allow me space. One more and the tiny boat might have capsized! The outboard was started and we chugged off downstream, under the new Severn road bridge, and the anchor was chucked over maybe 100yd west of the bridge, same distance from the shore. The men then simply jumped overboard, one at a time, ending up chest deep in the water. That was the first surprise as I hadn't realised just how shallow it was.

Like so many traditional ways of fishing, no one knows when the first users of the lave-net set foot into the Severn. And 'set foot' is by no means an understatement for that is how, for generations stretching way back, local men from the Portskewett, Caldicot and Magor areas of South Wales fished here by standing (or indeed walking around) waist deep in the river with their special nets. What is known is that, in about 1919, with men returning from the battlefields of northern France, the leaseholder of the Black Rock fishery agreed to rent it out to a handful of them. Of course, the lave-net is used over much of the Severn, but here I refer to this area, the last proponents of its use.

The lave-net is a hand-held net, 'Y'-shaped, used to catch salmon and sea trout in various locations along the Severn, the Black Rock today being the one lowest down the river. Both Deryck Huby and Don Riddle kept licences for lave-nets, which they sometimes fished if they reckoned fish were about during the time they were emptying their putchers. Many other fishermen did the same thing, maintaining one lave-net. These folk often worked further upstream on the sandy seabed, and some even fitted sheep's horns to the tips of their lave-nets so they could run them over the sand.

The Black Rock fishermen, though, work over a very rocky bottom in what is regarded as one of the most dangerous rivers in the world. Thus there's a difference in the construction of their nets, which the fishermen make themselves. And I would like to think that the lave-net was the original such net and that it was expanded upon to be worked in deeper water as the stop-net.

To begin with, we will talk about the way a lave-net is constructed as that might help in describing the way it is used to fish, and is as Martin Morgan explained to me months after my outing with them.

'Firstly take a walk to Slade Woods above Rogiet in the early year when the sap is down and find a suitable piece of ash for the handle, which is known as the rock staff,' he started in his thorough description and the following is what I learnt.

Just over 2in in diameter will be just right after it's been debarked. The length ultimately depends on the height of the fisher, but a piece about 5ft can be cut. When completed, the rock staff should be about the same height as the top of the fisher's waders. The important thing is to get a branch that has a slight kick in it about a foot or more from the thicker end. This helps to balance the net, which is then easier on the arms when fishing.

Lave-netter standing on the rock that marks a parish boundary.

Ash is deemed the best because of its hard-wearing properties as the staff is also used as a walking stick when the fisher is in the water. Once the wood has had the bark removed, it is often left to dry unless it's needed straight away. On the thinner end of the rock staff it is squared off and tapered.

Next comes the head board, a flat piece of wood some 4in wide, 1in thick and about 18in in length, which acts as the spreader for the net. Pine is great for this, straight from the local timber yard, although anything can be used. Pine is easy to work and thus quick. A square hole is chiselled out in the centre to take the end of the rock staff. At one end the slot is made, a horseshoe-shaped open circle with one end longer than the other, and a chamfered edge on the bottom side, the main part of the opening. On the opposite end of the head board, a 'C'-shaped circle is cut out, about 2½in in diameter, with the chamfer on the top side of the board.

The rimes are what hold the net and one passes through the hole in the head board. These are lengths of willow from a pollarded tree, which produces a slightly bent branch, and they are 6ft in length and taper from about 2in in diameter down to about 1½in. Willow is regarded as being strong and light, although the rimes will break. Unlike the rock staff that can last for many years, the rimes are replaced frequently. Both rimes are bolted through the rock staff at their thicker end, about 6in below the head board, with the kick in the rock staff facing forward, away from the fisher. Thus, the rimes are hinged, and the other rime that doesn't pass through the head board can be engaged in the 'C' shape to open the net. To fold up, it is simply disengaged. Most wear occurs at the ends of rimes, at the point that the net joins, where it is dragged along the bottom.

The net itself is knitted by the fishermen. It's a horseshoe-shaped net and consists of fifty-six loops along its head line, which is its top edge. By law the head line should be 7ft 6in long, while the rimes are 6ft. Until the 1960s, nets were made from hemp twine and nowadays from polyester. It is said that the old timer 'Smacker' Williams could knit a net in a day and a half, although most say that it takes a lot longer! Generally nets last two seasons. It's a 4in mesh, made by looping the fifty-six loops over a brass bar that hangs down between two branches, two hooks in a ceiling or wherever they choose to work. The net is knitted using a template called the wand – a length of wood that has an exact circumference of 4in. The twine is wrapped round the wand, knitted through the loop above using a traditional net maker's needle, back round in a sheet bend and pulled tight against the wand.

Each mesh is made in this way, although for the first fifteen rows an extra mesh is added at each end, then fifteen rows of equal mesh length, followed by twenty rows where one or two are taken away each side, to end with thirty loops at the bottom.

The fifty-six loops of the net are then tied on to the head line, which is threaded through holes at the end of both rimes. The line then passes down each rime, through the mesh of the net, and is tied off below the head board. Lashings are made of the line to the rimes in about three places on each. The net is now ready, the only thing left being to cable-tie the green plastic licence plaque to the rock staff below the head board. Now we can resume fishing!

Before wading out into the river, or jumping overboard in this instance, several other points are necessary. Firstly, a basic understanding of the regulations. The *Cyfoeth Naruriol Cymru* or Natural Resources Wales would love to see the end of these traditional forms of fishing for salmon, as we've already discovered. They favour the riparian landowners and their ability to raise income by charging out fly fishing. But folk like the Black Rock fishermen persevere against increasing authoritarianism. When I was there they were allowed to fish with their lave-nets for three months during the summer (1 June to 31 August), and could only catch fifteen salmon in total over that time, and no more than five a month. For eight men, that's not much, but generally they do it to keep this way of life going, and presumably they enjoy it at the same time. In 2017, over the fishing season, they caught just five salmon! Indeed, over the last twenty-five years their annual average has been six salmon. Fishing can only take place in daylight hours and not at weekends. Tides have to be right too, and fishing is best on springs, and on the ebb too.

Next the fishermen must have knowledge of their grounds. Generally these lie south-west of Black Rock itself, somewhere just downstream of the Second Severn Crossing (officially the Prince of Wales Bridge). Their grounds have names such as 'The Gut', which is between the bridge and the red beacon, 'Gruggy' is where we (they, as I don't have a licence) were about to fish on the west side of the bridge. Then there's 'The Grandstand', 'The Monkey Tump' and 'The Sugar Cube' to add to the evocative list of names, and basically these are areas of the seabed where they can move around.

All five lave-netters as the tide is nearly at the bottom.

Lave-netter standing in the water watching closely for signs of a fish.

The names originate from the times when the old fellows of the village sat around and talked about their favourite fishing spots. And, to be honest, there really aren't many topographical spots around the coast of the whole of Britain that the coastal fishermen haven't named. Maps are rare but I do know of two areas where fishermen have written down their memories: Bideford Bay and Kintyre.

Unbeknown to me, though I guess it's pretty obvious in hindsight, we had waited in the dinghy for the exact right moment to jump overboard. The tide had to drop to the right level – and remember the Severn has one of the largest tidal ranges in the world – until they could step out from the boat in their waders. It was slightly weird being anchored out in the midst of the Severn and seeing these blokes just climbing overboard into the river! Once in the river they must be able to watch for the signs of swimming salmon, know his patterns of swimming and be able to position themselves in the right place in the river. Then they have to manage the uneven seabed and find a place to stand with the net out in front of them, resting on or near the seabed. They must endeavour to place themselves in the position so that when the fish turns again, it will swim into the net. It's called 'cowering', this waiting. In actuality, it means standing in the water for two hours on the last of the ebb, watching and moving about, getting wet and tired, and hoping.

When that elusive salmon does come along into the net, they must recognise the signs, which can sometimes be very faint, and lift the net up, dig the end of the rock staff into the seabed to stabilise it, hold the net and manage to stun the fish with the knocker – the hefty lump of wood for this purpose. Only then can they thread a lanyard through the gills and the loop in the end and swing the fish over their shoulder to start fishing again. However, it has to be said, no one has caught two fish in one go in living memory!

Before the fish is taken ashore, a tag must be affixed. Fifteen of these tags are given to the group each year, and each specifies one of the allowable catch. Land without a tag and fishing would probably be brought to an abrupt halt if seen by a bailiff. Once ashore, a log also has to be filled in, giving the number of fishermen out, the times of fishing, state of the tide, and any catch. Oh, one last thing: they are not allowed to sell the fish, so it's cut up and shared among those fishing that day. I guess they just have to think themselves lucky.

Although their ten-year lease was renewed in 2018, an extra clause was inserted, giving the NRW the ability to withdraw the agreement at any time for any reason. In other words, cancel the lease at a moment's notice. In 2019, all they caught was two salmon, and, as usual, they hung the wooden salmon in the tree. But, in Martin's own words, 'We missed a few ... bit of a low effort with the threats from NRW, [it] sort of took the edge off it.' So, with the NRW not only happy about threatening closure, this had the added effort of discouraging those that were out keeping traditions alive. Then, in 2020 they stopped the catching of all salmon and they stopped lave-netting.

But the lave-netters aren't a gang of folk to easily give up. They petitioned and gathered thousands of signatures. They lobbied whoever they could, for the likes of Prince Charles had recognised their worth. After putting their case forward, the Welsh Senedd came out with the following:

> This Senedd:
> a) recognises the cultural and historic significance of the Black Rock Lave Net Fishery to Wales, across the UK and worldwide;
> b) applauds the work of the fishermen over years to engage with local communities, schools, events and media to spread knowledge and develop interest in the fishery;
> c) acknowledges the average catch of 6 salmon per season is insignificant in light of the residual mortality of catch and release in a salmon and trout rod fishery; and
> d) calls on Natural Resources Wales to accept the distinct nature, cultural and historic importance of the fishery and adapt its policies and regulation.

At the time of writing nothing has changed, which, in many minds, including mine, is an abuse of power. Only time will tell. Yet, for people who spend money on their licence and time on their fishing gear to keep this tradition alive, it really is a travesty that these bureaucrats who really have no clue as to what they are doing, but who only listen to the loudest voice, can, at the stroke of their pens, just delete another generations-old custom from the foreshore. Meanwhile, anglers have carte blanche if they choose to ignore levels of salmon stock and 'catch and release'.

The anglers' 2021 season started on 1 February and ran until 7 October. They have an unlimited number of anglers, an unlimited catch, no legal requirement to record their catch or effort, a season nearly three times longer than that of the lave-netters and none of the habitat's regulation assessment to bother with. Thus this one fishery will indirectly (depending on several issues) 'kill' more salmon in one season than the lave-netters have taken in decades. Yet they are still without a licence to fish and can only spend their hours on the low tide fishing grounds unearthing ancient fish traps and redundant Second World War munitions. How dare these bureaucrats simply wash this centuries-old proud heritage down the drain.

In order to begin lave-netting again, the fishermen then offered to reduce their permitted catch from fifteen to five salmon and to become a scientific fishery in regard to submitting data of their fishing. They awaited a reply.

When that reply came, it was a definite no from NRW, thus ignoring the statement from the Senedd. For 2021, no lave-net fishing occurred throughout the short season and the visitor centre never opened. Such a massive injustice for these few fellows who previously had persevered through thick and thin to promote, educate and continue a respectful tradition.

One final thought: we've looked at the stop-net, and soon we will learn of its smaller sister, the compass-net, and the suggestion is that these two stemmed from the lave-net, with fishermen determined to strike out in a bigger way. The technique is almost the same, with the stop-net fishing from a boat in deeper water, and a stronger current. The compass, as we shall see, is simply the stop-net in shallower waters. But in my mind, they all come back to the humble lave-net, a device made simply from the forest and fields – hazel, pine, flax, hemp – hand-held and adaptable. How fitting then that, possibly being the first 'device' to fish for salmon, they are now the last to leave the river. Please don't shut the door behind them, for then there's always the chance to return.

12

FISHING WITH A CORACLE

Having passed over the border into Wales before visiting Black Rock it seemed quite a jump to Carmarthen, where I'd arranged to go out with the coracle fishermen of the River Towy.

Now no one knows who gifted the coracles to the rivers of South Wales, or when they did. Old they are and it is generally believed that one of the early forms of sea transport was the hide-covered vessel built upon a frame of willow and hazel. It makes sense really. Why throw away a useful commodity after you've skinned and roasted your Sunday lunch? Both St Brendan and St Columba are said to have moved about aboard currachs, larger versions of the coracle. Gerald of Wales wrote of them too, as did many other lesser-known writers. And obviously some of these types of craft were used for fishing. In Wales they were called coracles and have been used to fish many of the rivers for salmon and sewin (sea trout).

Coracle fishing was undertaken in pairs, two coracles floating down stream with the current with a 40ft-long by 18in-deep net suspended between both vessels. One hand on the net, the other controlling the vessel with the paddle. This form of fishing is said to be the oldest form of trawling and was generally frowned upon in the nineteenth century, when those doing it were regarded as poachers. Others would describe them as working-class people out to catch some dinner. In a 1863 government report, coracle men were described as 'often lawless and always aggressive ... difficult to detect and almost impossible to capture'. This might hint to the fact that, without stating the obvious, in these rural settings, as in the salmon rivers of Scotland, the riparian fishers had more clout politically then as they do now. Not surprising, these days it is only the rivers Towy, Taf and Teifi that have been home to coracle fishermen since the Fisheries Act of 1923, when most of the licences were withdrawn and fishing stopped in the Usk, Severn and Dee. Again, those who fish today do it more to retain the heritage than make a profit and the tradition is to pass licences down through the generations from father to son, though this doesn't always fit individual cases.

Back in 2016, I spent a couple of evenings with a few of the coracle men at Carmarthen – Andrew Davies, Malcolm Rees (whose father Raymond I will come to) and Dai Elias, one of the older generation of fishermen that fish the Towy. I met Andrew where they keep their coracles by the old heritage centre on the quay at Carmarthen and he showed me his net, which is a bit like a trammel net with two layers of netting, one some 4in in mesh the other much larger, with floats atop and weights below, and running on metal rings. These used to be cut from cows' horns and the rope from the hair from cows' tails, though today they use man-made fibres. The amount of weight depends on the speed of the current on the river, a faster flow needing more lead. Due to the nature of coracle fishing, I was unable to join them on the water and had to simply observe and photograph from the riverbank, walking along it at the speed of the current. However, in the fading dark as dusk approached I really didn't note the number of stars at any one time, the reason for which will soon become apparent!

Coming down the Towy. The town bridge is just visible underneath the newer modern road bridge.

The coracle men drift down from the town quay just below the older of the road bridges and can go some 4 miles downstream in the tidal part of the river. There were eight pairs of fishermen licensed to fish and when they all do so, they must keep 200yd between each other. The first evening there were two pairs, including Andrew Davies, and they set off in the dark, at a time when my camera decided to play up. I would have had some terrific shots if it hadn't! Second time round, the following evening, there was what they call a 'drift tide', when the river is running very slowly, so that I was able to keep up using the path alongside the river. I followed Malcolm and Dai in their short run of little over a mile and they netted three sewin. They were made up and shouted over that I brought them luck! That made a change after being called a Jonah on more than one occasion! Fishermen often blame others for poor fishing,

It is said that there were 200 pairs fishing the river a century or so ago and it's hard to imagine how so many survived and caught fish. In 2016 the season lasted from 1 March to the end of July, though the salmon itself can only be caught between 1 June and 31 July. The Taf runs through the same season, while the Teifi men get an extra month up to the end of August. Not only are the seasons constantly squeezed, as we've seen already, but the licence fees rise and weekend fishing is banned altogether. The authorities would in general like to see the end of these various ancient forms of fishing, though the diehards refuse to give up. In 2019, the Welsh government decided to make all salmon 'catch and release', as we saw at Black Rock. Furthermore, the coracle season for the Towy/Carmarthen was reduced from five months to three (previously 1 March to 31 July, so as of 2020 it is 1 May to 31 July). The Carmarthen Coracle and Netsmen's Association have always abided by the strict regulations set out by NRW. They say that they can only hope NRW police and enforce even stricter regulations, penalties and legal action on those who continue to pollute our rivers, although there is doubt that they actually do. Furthermore they state that:

the restrictions implemented on all river fishermen are understandable. However, until issues are resolved with commercial trawling on fish feeding grounds and agricultural and industrial pollution on the rivers and tributaries, the restrictions will have little impact. Needless to say, we are also deeply concerned what effect this will have on our heritage. Coracle fishing will no longer become viable as each year we have fewer participants to ensure our legacy survives. We agree that our Welsh rivers are in a worryingly poor condition and fast, appropriate action needs to be taken before fish stock vanishes completely. With two slurry spills in the River Towy over one week alone in 2019, this doesn't bode well for the environment and little wonder why the river faces its lowest ever stock of fish.

While the authorities are always quick to judge commercial fishermen, it seems the antics of farmers and industry are quickly overlooked in many cases. In the middle of last year (August 2020) we heard about another polluting spillage into the Wye, though the NRW seemed to drag their feet in even finding out the cause, never mind in taking action.

Although not quite as hectic as the Towy, there were supposedly some 300 coracles on the Tefi in the nineteenth century (i.e. 150 pairs) and now there are twelve licences, while the Taf has just one surviving pair licensed. We were into the final month and these Carmarthen men were hoping for a bit of salmon instead of the mullet and sewin they'd been catching. Salmon, it appears, were few and far between. So was the sewin! But then, that's not anything new.

No talk about coracle fishing can omit a mention of Raymond Rees, ex-coracle fisherman and ex-fishmonger, although I did allude to him by name above. When I was living in Carmarthen in my house in Spilman Street, the River Towy was almost in view from the upper floor where I slept. Raymond had his fisherman's store just down the road and, on passing with Mono the dog, we'd often have a chat. Sometimes we passed the time of day and at others, we engaged in extended discussions about his time on the river. I remember one such mid-morning time back in about 2002, Raymond told me, 'See, to be good on the river you have to understand the *clefwcwr*. That's hard to put into English. It's a time, not fixed, see, but moving.' Like the River Towy, the edge of which we're standing on, where he and his ancestors had been fishing for generations.

'Back to 1710 they were here on this bit of the river.' He pointed to the water, gently flowing under the town's bridge that was built in the 1950s to replace the earlier one, the first road crossing over the river until the bypass was built in the 1970s. 'Two miles upstream and seven down, that's where we can fish.'

I asked him to try and explain *clefwcwr*. As far as I could understand, it is the time between dusk and the moment in time when the river stops flowing; when the flood equals its normal stream. Sometimes this lasts half an

Raymond Rees showing how a coracle should be carried on the back with the leather strap around the chest. The knocker is in its own strap and the paddle secured through his arm, which prevents the coracle hitting the lower back.

hour and at other times maybe four hours. Dusk, he said, is, by agreement, the time when seven stars are visible in the sky. Assuming there's no cloud that is. I didn't ask what happened then!

One thing that became clear after spending a couple of hours with Raymond in his shed alongside the river was that coracle fishing wasn't easy. There might now be a resurgence of coracle building around Wales – and further afield for that matter – but using them was altogether a different skill.

'Success is down to reading the river,' he continued. 'Don't get it right and you might as well fish in the bath.' He pointed to a block in the nearby bridge and told me how he watches that. From the height of the water, and the pattern of flow around the base, he can tell the time and place to fish.

Up in the roof of his workshop, with a mix of woodworking equipment, strips of timber, netting and various pots of whatever, there's an old coracle.

'Built by my grandfather over fifty years ago,' he said. We pulled it down. 'Now our coracle building here on the Tywi is truly a work of art. We call it *eielgody*, meaning to resurrect.'

He expanded on this by showing me how the coracle consisted of seven horizontal, seven vertical and two diagonal strips of ash. The gunwale consisted of several rows of bent hazel – or willow, he added – the first one starting on the left side, the second on the right, around the front of the boat. Next they come around the tail – the stern – of the boat, and so on, building up a series of plaits. Then the calico shell had been laced on. Above this, a final plait of four strands of hazel formed the top of the gunwale. As this is the first part of the vessel to rot, it can easily be removed and replaced – resurrected so to speak.

'Innovation,' he said. 'No other coracle has that. That's what sets us coracle men apart from others.'

Fitting the priest – or *cnocer* – into its leather strap, he lifted the coracle onto his back, suspending it over his shoulders with another leather strap. This, he assured me, was another difference, as normally this strap would be made from several strands of hazel. To walk with the coracle, the paddle was used as a lever under the arm to lift it away from the back of the legs.

'This ensures the legs don't get wet.'

Another innovation. And another was the fish box beneath the seat. Below this was the *orlais* – a sort of bulkhead, to support the seat (*astell*). It had the added advantages of preventing the fish slithering to the tail of the coracle, where it was possible one might jump out. When carrying the craft, it served as a box to retain the fish for the walk home. The technique of lowering the coracle back to the ground was also exacting to ensure the fish remained in the coracle and were not dumped onto the ground.

'Yea,' he continued, 'we coracle fishermen evolved our coracles much more than upon any of the other rivers. Not only that, we were the early warners of any peculiarities on the river.'

The Towy fishermen were simply the best, he assured me. So I asked him what about the Teifi coracle men. He was dismissive; fishing there was free until the 1990s. Most of the Teifi coracles were built by one man in Cenarth, an ex-miner called Tomas, whereas all the fishermen on the Tywi built their own.

'Their coracles haven't really evolved like ours have because the tradition isn't there. Because here the ones that use them build them, we've been able to adapt to the needs of the river. As I said, see, success comes from an understanding of the river.'

Yes, I could see that as he stood by its flowing streams. Here was a man who had grown up with the river, probably from his first steps, a man for whom the river was still a mystery; yet few know it better. A man short in stature but strong in understanding every eddy, every ripple from a lifetime gazing into its ever-changing depths; and I can say I was proud to have met him, to be able to say I knew him. Sadly he has since passed on, but happily Malcolm is continuing this tradition on the river, though I think not from his father's Towy-side shed, which I did note at one time was for sale.

Before we depart these South Wales rivers, it's worth noting the wade-net of the River Taf, so-called because the fisherman wades into the river with his seine-net because the river is too shallow for a row boat. This used to be a two-man job, says Geraint Jenkins, writing in about 1970 but had, by that time, become a one-man undertaking. Today, of course, it's long gone. The ends of the 30ft long × 2yd deep net were attached to poles, one fixed to the shore (or, in the days of two men, by the second fellow) and the 'wader' took the other end out into the stream of the river, paying it out in a semicircle. Then he came back ashore to the fixed pole and hauled the net in with a pole in each hand. It was said by Jenkins that this was operated in three pools: Ferry House Pool, Cover Cliff Pool and Whaley Pool. Fishing occurred at night and the method was said not to be very effective due, in the main, to the strong current a man had to wade against, and in 1969 only ten salmon and five sewin were landed.

13

DRAWING A COMPASS IN THE RIVER CLEDDAU

From Carmarthen, we head west once again and it's 30 odd miles to the River Cleddau. Not long after being out with the Towy coraclers, early one morning I drove to Landshipping, a tiny hamlet in Pembrokeshire nestling on the eastern bank of the East river, just upstream of the point where the two rivers – East and West – converge. The confluence of these two small streams feeding off the hills to the north then becomes the Daugleddau that, along with other tidal tributaries, eventually ends up as Milford Haven.

Alun Lewis lives in the Big House at Landshipping, a huge home that he first fell in love with as a child. From the outside it looks more like a castle, twin turrets and all that. Inside it's more down to earth. When he came to own it with his then partner, Sarah Hoss, the rebuilding of it was featured in the BBC *Restoration Home* series, when the house was saved from complete decay and one half totally restored. Today Alun lives there alone unless his kids are with him, and earns a living off the river along with various other hats he wears as necessary. He's also the only compass-net fisherman who fishes in a traditional black-tarred wooden open boat, which has recently been restored.

Messrs Ormond and Edwards have attracted a degree of fame in the fishing circles of the Bristol Channel, yet both men remain unknown quantities and, although the actual date of their being seems unclear, theirs are names that are dragged along through history by a succession of authors, including me. It seems to be generally regarded that it was in the early 1800s that they moved from the Gloucestershire coal mines of the Forest of Dean to the anthracite mines of Pembrokeshire. Some say they worked at Hook and others at Landshipping, though in 1844, a time they might have been there, the latter was subject to Wales' worst mining disaster when the mine flooded and forty miners were killed.

With the Forest of Dean being in the proximity of both the Severn and Wye, it must have been either through part-time work or mere observation that they'd seen the stop-net fishermen of these two rivers. Now, with both Landshipping and Hook lying on the banks of the River Cleddau, I guess it wasn't surprising that, drawing on their experiences, they instigated a similar method of fishing on the upper reaches of the Cleddau. Fishing might not have been commercially exploited back then but it sure gave the participator a great bit of free protein.

Once Ormond and Edwards had adapted their system and presumably encouraged others to use a similar set-up – or maybe other fishermen and/or miners spotted them at work and thought theirs was a good idea – it became a widely used method on both the western and eastern arm of the upper Cleddau. The compass-net is in actuality a mere smaller sibling of the stop-net. It too resembles a geometric pair of compasses used to draw a circle and the rames (poles) were a bit shorter at 20ft in length. I'd seen various photographs of the net, as well as various examples of the smaller compass-net boats in previous visits to the river.

Alun Lewis in his small wooden boat that belonged to his grandfather. Here the net is down and bedded into the riverbed.

Alun is quite well known for his promoting of this fishery and as such has been interviewed and filmed widely. He's been born and bred into it – he hails from Llangwm a wee bit downstream and across the river – so should know it pretty well. Living almost on the river in the Big House, which dates back to the 1750s when it was possibly built as the coal agent's house, he feeds himself and his children from both the sea and the land. He shows me around, makes me a cup of coffee as I'm early and then we are off to do some fishing.

We use the slipway to launch his dory into the water; water that is instantly noted for its muddiness. The tide is dropping as we motor upstream, towing the K5 compass-net boat behind. This, he reckons, is possibly 130 years old. 'My grandfather had it built and he died in it when he was 84 and that was in 1966,' he says over the noise of the outboard. He added that it was a heart attack and the boat was found drifting next morning with his grandfather dead on the bottom boards.

The smaller boat weaves its way to and fro across our wake, seemingly eager to get to work. It was rebuilt a few years ago by Hugo Pettingfer at Alun's expense. At 13ft 10in it is slightly smaller than some of the others that now lie rotting, dotted about the Cleddau shoreline. I'd photographed various a decade or more ago, and even measured one up that I found on the mud below Hook village. Most were built, it is said, by the carpenters working in the naval dockyard at Pembroke Dock, the timber being smuggled out. Indeed, Alun found that much of the planking in his was in lengths of about 5ft, just about right to poke down your trousers and up your arm to walk – or limp – the short distance through the dockyard gate! One wonders whether the guards ever noticed the high frequency of a limping workforce. Nevertheless, eking the materials this way ensured minimum investment to procure the fishing ability. The net itself was very basic and wasn't large. It would have been made by the fishermen, while the wood for the rames would have been cut in the local woods. Furthermore, they are always painted inside and out in tar (again available in the dockyard), which cynics say is to avoid being spotted fishing at night. Alun used road tar, which was why, in the warmth of the day perched on the transom, I arrived home, the arse of my shorts covered in sticky bitumen! There were once, it is said, over 100 boats fishing, though today there are just six licences. Indeed, throughout Wales there are only forty-two licences for river fishing, which includes the Black Rock lave-net men, the coracle fishermen and a few seine-nets in Cardigan Bay. A sign for sure of the fortunes of fishing for pitted against them are 250,000 anglers.

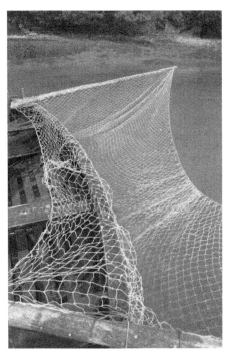

A closer inspection of the net in its up position.

We anchored the dory and waited a short while as the tide ebbed. Once a particular stake had appeared on the riverbank, we climbed aboard the smaller boat. It was about three hours after high water. One end of a long rope was attached to this and Alun rowed as fast as he could across the river, where a formidable current was still running, paying out the rope as we went. Once the end appeared with its hefty anchor, this he threw over the side and we returned to the other bank to find the rope in the murky water. This, then, was the 'chain' onto which we were attached so that the boat was broadside to the current, port side on. Alun swung the net over, fixing the spreaders to form the 'V' and then checked the depth of water. If the rames didn't dig into the mud, the whole net would swing over and upset the boat, throwing us into the water, and into the net, at which stage drowning was all but probable. I looked dubiously on as he lowered it. But he'd done this hundreds of times and could do it blindfolded, he said. Sometimes at night when all was black it was as if he was blindfolded. Finding the various stakes – there were thirteen around this part of the river – is indeed an art before the net goes in. Once down, it is propped up by a post with a crook at one end that one rame sits in.

Like the stop-net, the compass-net is a bag-net spread between the rames, the net streaming out below the boat. There were only three feeling strings attached to strategic points on the net but, as I found, these really do work.

'Have a go,' he said, passing me the rames and showing me how to hold the three strings between my fingers.

Proud as punch, I stood in the required position, leaning onto the cross-beam as I'd seen Alun do. Gotta get the right stance, I thought! The river swirled past and the sun shone down onto my neck, as I eagerly watched for any sign of a fish. I didn't see any but, after ten minutes, I felt the tiniest of vibrations along just one of the strings.

'It's like a tiny tingle, the smallest of movement?' I whispered. It was a question more than a statement.

'Knock out the prop and swing the net up,' Alun says back, leaving it to me as if it was second nature. It wasn't! But up came the rames and together we pulled in the net to find a fish. Wow! The cod end was untied to extract it, the knocker used to finish him off, the net tied up again and the whole apparatus set back into the water. Three times I felt that sudden and very slight tap on the strings and three times there was a fish in the net. Shame none were salmon!

On the fourth time there was much tapping and splashing so that, hoping for a whopper, I passed the net to Alun but when he brought the net up, alas, there was nothing. He reckoned it was probably a fish passing through the mesh or even possibly a fish on the backside of the net giving it a flick of his tail to tease us. On another occasion, on feeling a vibration, there were two plastic mugs in the net!

Conditions were perfect. With an osprey hovering over the trees of the wonderfully evocatively named Peepout Wood, the sun shining down and glassy waters being disturbed by the occasional ripples of a slight breeze, we chatted, ate and spent time in silence in the olden way of fishing, for fishermen used to believe the salmon could hear their voices. Maybe that was why there were no salmon, though many accuse the shortened season that lasts from June to August, with fishing prohibited at weekends and Monday mornings. But even when basking in the idyllic, it was easy to see how different it would be with a chop and a wind. In those conditions, putting down and bringing up the net can be dangerous times when the small boat is plunging and swinging wildly against the ebbing tide.

We fished for about three and a half hours, right down to low water, where we were only in 2ft of river. It's at this time that Alun has caught many a salmon but the fish were not to avail themselves. The current changed direction in a matter of minutes and Alun brought the net in and hauled in the rope, while I unfixed it from the stake. As we motored back, he pointed out various other stakes in the riverbed, all of which uncovered about half tide. To fish in any other stage of the tide was a waste of time, he said.

When he's at home – he drives trucks to supplement an income as well as operating the fishing charter boat *Cleddau King* – he has the whole of the upper Eastern Cleddau as his work ground. Whether it's fishing, shooting birds, hunting rabbits or foraging for wild plants and berries, he's happiest outdoors, living the land; living the dream, so to speak. The house is an obvious financial burden, given that it was derelict when he and his ex-partner first bought it, but this was their dream and one that has progressed so far. Situated in the Pembrokeshire National Park alongside what is considered to be the secret waterway, many have described the compass-net as a romantic way of fishing. But I beg to differ. It tends to catch one fish at a time, thereby being a rare, selective way of fishing while, through seasonal controls, its catch is very limited over that season. It might be an old way of fishing but, in this day of overfishing, it has become one of the most sustainable to fish. Long may folk like Alun Lewis continue to keep these traditions alive, for their survival beyond his generation is a most unlikely affair. And that is, indeed, sad. No wonder the compass-net has been referred to as 'the Gift of Ormond and Edwards'!

14

THE CLEDDAU KING

Once the compass-net season is over, Alun must turn his attention to other matters. One is the local herring fishery, which is a winter preoccupation and thus, several months after being out on the water with him, I was back. Late winter, with the March equinox just around the corner; it was a typical grey nondescript sort of day, a threatening-to-rain-yet-didn't one, with little warmth, and a never-quite-cold type of one. Nevertheless, there was a bit of breeze whipping over the surface of the water, causing small ripples, as we arrived at the water's edge. Entering between the stone gateposts of Alun's residence, we were met by his yellow JCB reversing down the driveway with the compass-net boat, K5, hanging from the rear activator bucket, heading toward the slipway. It appeared the small boat was being dragged away from its winter slumber.

But early awakenings are always vital when the tide needs catching. Before long we – Alun, his children Freya and Geraint, my son Otis and myself – were heading downstream, aboard the *Cleddau King*, Alun's charter boat, trailing K5 behind in the wake. As she streamed across our wake, heading from side to side, she conveyed to me the excitement at once again living up to her expectations as a 130-year-old work boat. No doubt Alun's grandfather had also moved downstream at this time of year in K5, although he'd have been rowing the boat rather than being towed behind another with a thumping great big engine below the aft deck.

Heading into the Daugleddau, we motored another 4 miles to where the river narrows somewhat, just short of Lawrenny, and anchored in the middle of the river. With nets already stowed aboard, we climbed over the side and into the small boat, all except Freya, who we left aboard with the camera!

Black-tarred boat complete with the net in the stern. The other boat is the one that caught 3 stone of herring.

With me on the oars, we moved towards the right-hand bank before Alun started shooting the net over the stern. First the stone weights go over, three set at one end of the footrope, fixed by tying ropes around the stones, along with a buoy attached to the top rope. Then the nets, all four of them, 50yd long each one with smaller stones attached at intervals, before finally the weights and buoy of the other end.

'The stones are grandad's,' he tells me when I ask about them.

'As old as the boat then maybe,' I suggest. He looks up and smiles.

The nets are set at a depth below the surface and generally he fishes half an hour before low water and for an hour over the slack water, so that it can be hauled before the onslaught of the flood. With a 25ft tidal range just around the corner at Milford Haven, the tide can tangle a net in minutes when it wants.

Sometimes he fishes during the day, sometimes at night, when the herring tend to come up to the surface to feed.

'Dawning tides were usually reckoned to be the best,' he said when I asked about the size of catches. 'The best catch I've had? Once there was about 125st in one haul and I spent the best part of the night picking them out of the net.'

I rowed back to the *Cleddau King* and embarked Alun and Geraint before taking Otis off for a row around. I pointed out the sounds of the riverbank to Otis, who seemed amused that we were aimlessly rowing around in big circles. A statue-like heron was watching us, hawk-eyed to the slightest movement in the water but possibly disturbed by the splash of the oars that seemed to echo around the trees that lined the river. Between tree line and water, the mud of the bank suddenly glistened when a ray of sun escaped from between grey cloud, not to last more than a few seconds before being stolen away. The dampness of late winter sent a shiver through me, even though I was wearing thick jumpers and oilskins. A plop nearby alerted us to the presence of fish and even the heron turned its head. Before long, a hail from the mothership suggested it was time to haul up the net.

But I was wrong. Alun had boiled the kettle and so, once back aboard, we wiled away some more time by drinking copious amounts of coffee, the sun not appearing again and the cold seemingly on the increase. When swirls around in the water suggested the tide was beginning to turn, once again Alun and I hopped into K5 and I rowed over to where the first buoy was. Hauling commenced. The initial tangible excitement was somewhat subdued when the first net appeared empty, then a silver flash beneath the

Alun hauling in the net with one of the three herrings!

water and two herrings appeared. By the time we'd brought the whole lot in, we'd realised three small herrings. Three bloody herrings. It was hardly worth the effort, never mind the fuel used running the few miles down from Landshipping. A monumental disaster of a fishery and not for the first time did I wonder just whether I was a Jonah. I quickly remembered all the other times when I'd been fishing and we'd landed some relatively massive hauls.

Like the time, back on a crisp January morning in 2012, when I was stow-netting in the River Blackwater in Essex, aboard Paul Winter's lovely 47ft sailing smack *Maria*, CK21, the stow-net being an ancient form of fishing that involves placing a net beneath a sailing smack by means of a complicated assemble of wooden beams, ropes and nets that are lowered over the bow of the anchored vessel.

Without being boring, it works a bit like this: first the boat needs a hefty windlass and strong anchor chain. The two massive baulks of timber that form the mouth of the net are heaved overboard, using the wooden baulk-davit on the starboard side of the stem. Just abaft of this is the windlass used to set and adjust the anchor chain and the wink chain over the davit. The system of the stowboat netting works thus: the two baulks sit above and below the mouth, from which the net is suspended. A line from each end of the baulks – called a hand fleet – lead to a bridle rope, which is shackled onto the anchor chain so that, as the chain is lowered, the gear goes down with it. The upper baulk has a line each end – a templine or hip rope – leading onto the deck, which controls the angle of the baulk. When the gear is at the required depth, the wink chain is let go, which in turn lets out the lower baulk, thus opening out the mouth of the net. The net itself streams out beneath the smack and the end is marked with a buoy. We took half an hour to get the gear down while smacks men were said to do it in three minutes!

After an hour of rest in the small cabin, chatting over mugs of steaming hot tea, the condensation almost freezing on the deckhead, we ventured on deck to haul in. Immediately we hauled in the anchor chain and it was pretty obvious there was a good haul in the net. After plenty of sweat and some swearing, the baulks appeared on the surface and the net streamed aft. This we pulled in inch by inch, removing sprats from the mesh as we progressed. Once the cod end was visible we saw to everyone's amazement that it was full. Getting a strop around the net, set through a pulley off the mast, it was impossible to haul up because of the amount of fish. We just didn't have enough muscle, even though the boat heeled over. The only answer was to split the net and let some fish go. Then we were able to haul it over the gunwale and let go. Sprats and small Blackwater herrings flooded onto the deck, even some cod, whiting and codling, which just shows how efficient a marine trap this was. Covered in scales and warmed by the exercise, we soon had the deck cleared and gear aboard, before raising sail and returning to the boat's mooring in the Pyefleet Channel, and to ship the catch ashore.

Blackwater and Cleddau herring are very similar and maybe cousins in some weird way. They are small herring, as the Clovelly are, but are different in that they are anadromous, otherwise called river herring; in other words they spend most of their lives in the sea but return to rivers to spawn. But that didn't help us as we stared at our three little herrings. With the net stowed and K5 streaming astern, we motored back upstream. A couple of hundred yards along the river we passed a small fibreglass boat with a red-clad figure who was just beginning to haul in his nets. Maybe he wasn't trying to shield them from us, maybe he was, but I couldn't see any herring. Nevertheless, by the time we had moored up the *Cleddau King* and all got safely back onshore and K5 happily resting in the garden once again, the news had filtered through: that red-clad fisherman had caught 3st of herring. It just showed what a difference a couple of hundred yards can make!

15

FISH WEIRS (AND A FEW MORE CORACLES) WHILE RUSHING UP THE WELSH COAST

From Landshipping it's a short hop westward to St Bride's Bay, a delightful horseshoe-shaped bay with various hamlets or 'havens' as they are called (Martin's Haven, St Bride's Haven, Little Haven, Broad Haven and Nolton Haven) before arriving at Solva and then Porthclais. All these small coves and beaches were once home to fishermen. At St Bride's Haven, a tiny chapel allowed the men to pray for successful catches of herring. Because the fishing was good, a salt-house was built that obviously upset some higher beings for, as the local distich went:

> When St Bride's Chapel a salt-house was made,
> St Bride's lost the herring trade.

Once past the mini-city of St Davids, it's a wonderful drive along the edge of Cardigan Bay, when the road allows proximity to it. Again the little coves were centres of local fishing or, in the cases of Abereiddi and Porthgain, the export of slate from the nearby quarries and shipbuilding at Abercastle. Indeed, a refreshment call at the Sloop Inn at Porthgain, dating from 1743, is a must and is open throughout most of the year.

The road arrives at Fishguard, stepping off point for those heading to Rosslare in Ireland, some 54 miles a bit north of west from here, across St George's Channel. Well, to be exact, the ferry port is in Goodwick, or *Gdig* in Welsh, and not Fishguard, which lies a mile further along the main road. But Goodwick we are heading for to view the remains of the Goodwick fish weir.

Fishguard gets its anglicised name from the Old Norse *fiskr-gardr*, which equates to 'fish yard' or 'fish weir', depending on your train of thought. In Welsh it's *Abergwaun*, 'the mouth of the river Gwaun', which makes more sense. So, in confusion, we have the fish weir across from Fishguard, which makes no sense at all. Nevertheless, the Ordnance Survey maps show it, and when I was sloshing around in knee-deep water following its path, it was clearly visible, although of course only by the line of stones. Access, I found, is best via Quay Road and across the footbridge, away from the ferry traffic.

Now, we have already come across fish weirs in Minehead, several chapters back. In reality, I could rattle on about these ancient structures for pages as I've spent more than twenty years studying the various types all along the west coast and beyond. I've an overflowing file with articles, notes and drawings of the layout of a host of them, from Porlock Weir right up to Gairloch, way up north. But for now, we'll talk about some of the Welsh ones.

A small boy with two lobsters standing in the fish weir (gorad bach) by the old lifeboat station near Beaumaris, Anglesey. (Bridget Dempsey)

Fish weirs are one of man's earliest forms of fishing, back in the days when he was the hunter-gatherer, roaming the coast in search of food. He lived off the land and the sea or rivers. That in itself was a dangerous occupation, given the nature of life back in those days. In Britain, the landscape was one of a heavily forested hinterland with prime rivers for catching fish in many parts. In others it was the sea, which again was rich in all manner of fish and seafood. If one considers what was available before man started a determined effort to destroy the fruits of the sea, the foreshore alone gave him shellfish – mussels, cockles, oysters, whelks, winkles, limpets and other shellfish – while the rock pools might have contained small fish, crabs and shrimps. Then the rivers were rich, with their salmon, trout, sea-trout, eels and flounders, while just offshore were all manner of shoaling and bottom-feeding fish and other shellfish such as lobsters. Even the inland lakes produced fish such as pike and tench. What a banquet could be collected in a short time!

I might have given a false picture here that the life of these hunter-gatherers was a simple one, although the truth is the opposite. It was a strenuous and hard life, with the added conflict between tribes and different roving bandits who were always keen for a free meal and not concerned about the value of human life. History is full of tales of death and destruction, and the fisherman was not exempt. However, as time went by, his techniques of fishing must have improved. His fishing without any gear – just his hands and feet – must have taught him the use of fishing tools such as knives, spears and long-handled hooks. Divers used such hooks for loosening shells and hooking octopus. Over the centuries all manner of methods have been used for catching fish. These include the use of animals such as dogs, otters, cormorants, turtles, octopi and porpoises, using mechanical ways of stupefying fish such as dynamite, toxic plants, clamps and rakes, and electrical fishing. Shooting, spearing and harpooning were ways of capturing fish, especially after it was discovered that light attracted fish.

No doubt by studying the topographical layout of beaches he learnt the art of damming the beach to catch fish, for fish weirs are just that: barriers that create pools once the tide has receded. That is also why they are generally referred to as 'fixed engines'.

Documented evidence of weirs is scarce, though three traps excavated at the Late Kongemose site Agerod V in southern Sweden are said to date back more than 6,000 years. Another, at the Ertebolle site of Jonstorp, still contained a cod and several have relatively recently been excavated in Denmark that date from the Mesolithic and Neolithic times. In Britain, various fishing baskets and fish traps at Goldcliff, in the Severn estuary, have been observed in the minerogenic sediments and date to around 4000–5400 cal BC. Many others have been established for hundreds of years. Indeed, as Davis recounts, 'the first settlers in Queensland, some of whom lived much among the then Blackfellows, have left very full descriptions of the stone-weirs used by these primitive people who were still in the Stone Age of culture'. The same he says of the Native Americans of Virginia, who developed 'great fish weirs and fish pounds'. Davis also suggests that it is not improbable that the earliest method of fishing in any quantity was by stopping the mouths of narrow tidal creeks with brushwood or stones but when it was discovered that this also blocked the ingress of fish, the weir was developed by forming an opening that could be closed on the ebb. Davis gives the 'Fish-ponds' of south Devon as an example. In the Outer Hebrides, 'one method of catching fish which was once common was by building yares or stone dykes across a river estuary. The fish that swim in at high tide when the yare was submerged were left stranded when the tide ebbed, and could be collected without much trouble.'

As has been mentioned previously, fish weirs are only capable of working effectively in tidal waters and, as so mentioned, Britain has the remains of many littered about its coast, especially on the western side of Scotland, England and Wales. These weirs were positioned in such a way that they used the natural swimming behaviour of the fish, which, in shallow flowing water, tend to swim parallel to the coastline. On the flood many tend to swim towards the shore and away on the ebb. Thus, as the tide recedes, the entrance should be behind the fish. They were generally constructed with stout oak posts embedded into the foreshore, with stone walls built up for the first foot or two, and then hazel or willow was used to weave a framework that allowed a flow of seawater to pass. The weave was more tightly woven at the bottom, thus preventing fish escaping, and looser at the top, ensuring the greater flow as the higher level of tide passed through without damaging the structure, for the pressure of water can be immense. Some weirs had intricate openings at their mouths with sluices so that these mouths could be opened or shut. Those Minehead ones are good examples.

John Girling (senior) with a landing net full of whitebait in the gorad bach weir. (Bridget Dempsey)

Weirs follow various patterns and an attempt has been made to classify them as to whether they are active or passive. Again the shape is important as a couple on the west coast of Wales are crescent shaped – described by one writer as 'somewhat shaped as a boomerang'. Others are 'V'-shaped, often with several side by side and a net of some form across the neck where the fish are caught. But in reality they come in all shapes and sizes. As I was writing this, the family were watching a programme on television and suddenly there was a mention of fish traps. I stopped and listened to hear how dolphins were trapping fish in a manoeuvring technique similar in some ways to these structures. And a very successful technique it appeared as fish after fish flopped into their jaws. Maybe, then, the development of fish weirs simply came about by watching closely the ways of the creatures that inhabit the seas.

Thus this Goodwick weir is shaped like a hook. The remains, a pile of large boulders, run out from the shore at an angle of about 20 degrees, away from the seaward side, for some 100yd and then the hook returns shoreward for about 20yd. There's a vague sign of some more stones slightly further east, which could have been another arm. It must be taken into account that the foreshore has been rock infilled as part of the embankment over which the access road to the ferry port runs. Nevertheless, the general shape of the one arm can be plainly seen, though, in the shallow water, I didn't find any signs of stumps of wooden posts.

We drove north once again after a brief visit to the small tidal harbour of Fishguard, where a straight line of stone suggests the existence of another weir at some time. There's a four-storey former herring and pilchard curing factory still standing, although obviously not being used for any fish activities, which dates from the time when both fish were landed here. Pilchards were the domain of Cornwall but in the eighteenth century merchant and shipowner Samuel Fenton developed the harbour and introduced pilchard curing to export to Italy and further afield. I'm not sure it was a very successful enterprise but at least the building survived. Today it is home to the Fishguard Sea Cadets.

Newport Sands was once the base for seine-netting the river entrance for salmon using 19ft-long black-tarred open boats. Several fishing stations (*ergyd*) were licensed and each had five men to work them. Generally they wouldn't start fishing until Y Garreg Fach showed above an ebbing tide and if the weather was adverse, they'd work the river. Coracles worked further up the river.

Cardigan lies another 10 miles along the coast, beside the river Teifi, from where it gets its Welsh name Aberteifi. St Dogmaels, on the south bank, slightly further downstream, was the base of the local fishermen, who kept their boats at an area called the Pinog. Out in the bay, two fish weirs are known to have been worked, one between Cei bach and Tywyn Garreg-ddu and another at Traethgwyn, though I could find no physical signs of them.

The Teifi was a renowned salmon river and two particular spots were licensed coracle fishing spots: at Cilgerran and Cenarth. Cenarth also is the home of the National Coracle Centre, which is situated close to the Cenarth Falls where the salmon sometimes leap. The museum was closed when I was last there, and I'm unsure as to the opening hours during the season, although appointments can be booked at most times of the year. It is situated in the grounds of the seventeenth-century mill, which is also open to the public at times. The museum has a fantastic collection of coracles from all over the world, not just England and Wales. Some coracles are still licensed to fish the river, as they are in Carmarthen.

The 'chiefest weare of all Wales', according to George Owen of Henllys, in his 1603 *A Description of Pembrokeshire*, is situated on the river in the gorge below the castle at Cilgerran. This has been in existence since the Middle Ages and records show that, in 1314, permission was given for its reconstruction after it had affected the carriage of stone and timber for the construction of Cardigan Castle. Again Owen writes that it was 'built of strong tymber frames and artificiallie wrought therein with stone, crossinge the whole river from side to side and having six slaughter places, wherein the fish entering remain inclosed and are therein killed with an iron crooke proper for that use'. He added that 140 salmon were once caught in the weir in one day, that they were of the 'most excellent, and for fattenes and sweetenes exceeding those of other ryvers'.

Heading north once again, the road skirts round the coastal villages of Aberporth, Tresaith and Llangrannog, each home to fleets of herring boats up till the middle of the twentieth century and now the domain of holiday homes and tourism. Past Newquay and Aberaeron, with its laid-out harbour and town, to Aberarth, which is little more than a hamlet on either side of the River Arth. However, it is possible to discern the remains of eight weirs in the vicinity of the village. All are fairly indistinct, although an outline can be traced. I measured them back between 1999 and 2002 and all appear to be crescent-shaped or similar, out from the foreshore, and are close to today's low-water mark. At Llanon two weirs can clearly be seen at low water. After Aberystwyth the main road moves inland, away from the coast, and it's a fast run up to the north coast. The remains of other weirs must lie in the shallows and one of the oldest was documented off the north coast of the Llŷn peninsula, but, in the hope of not boring readers, we'll head over the bridge to Anglesey.

Anglesey has its own weirs. Red Wharf Bay, Lligwy and Beddmanarch Bay, at the mouth of the River Alaw, have examples. Various examples exist on the Menai Strait and they offer the best chance of understanding how these structures work.

The *gorad bach* – literally little weir – on the east end of the Menai Strait has a 'bass trap', which is a little sluice that can be opened to let whitebait out (a common fish in these parts). Bigger fish such as salmon and bass tended to linger outside awaiting a tasty meal, although the owners of the weir, John and Wilf Girling, would also linger above with a lap-net to catch the unsuspecting salmon or bass. This weir, along with another larger one a mile to the east called the Trecastell weir, date back several centuries and were very effective at catching all manner of fish: herring, whitebait, salmon, mackerel, bass and even the green-boned garfish. They were the same shape and consisted of a wall running at right-angles from the shore, and out to the low-water mark, and then another along the low-water mark perpendicular to the first. At the right-hand end, looking from the shore, the end that faces the ebb, a short wall runs back on itself at a sharp angle, thus forming the 'crew', where the fish are unable to escape. The *gorad bach* was in use until the 1960s and the owners then only packed up as it was obvious that the seagulls and bait diggers were having more of the catch than they were! One weir in this area, called the lyme-kiln, was leased to Thomas Norrey in 1438 for twenty years, while Thomas Sherwin paid a rent of 6*d* a year ten years later for another said to be 'lying between the lyme-kiln fishery and the house of the Friar Minor of Llanfaes'. Given that the *gored bach* is in the vicinity of Llanfaes, it could be assumed that the report refers to this weir.

The Menai Strait had several others of these fish weirs and one was restored in recent times. Current regulations ban their use, so this one has holes in the restored wall that allow any fish to escape. Situated on the small island between the two bridges over the Menai Strait, the weir can be studied from above and if the tide is high, its shape can clearly be seen. Another used to be worked on the island – the remains are still visible – and the remains of two others can be found nearby on the Anglesey shore. One of these two uses the method of building a barrier between two islands so that the fish can pass around the outside of the island and towards the shore on the flood, but are then are caught when trying to make their way between the islands on the ebb. Further north-east, the outline of another is clearly visible at Cadnant in the mud at low water. Another used to lie across the Straits on the Bangor side, a vague remnant still visible with the tide out. Another can be seen at the mouth of the River Ogwen.

Some miles east are the remains of two fish weirs at Rhos Fynach – the 'Monks' promontory' – which date back to the twelfth century. The western weir worked up to the time of the First World War. In the 1860s, a celebrated dog fisher was presented with a solid silver collar by public subscription. The dog, it seems, was excellent at catching salmon and thus spent his days at the weir.

Another few miles east, the River Clwyd was once renowned for its salmon. The ancient hamlet of Foryd was once the port of Rhuddlan but it now mingled in with Rhyl. Between the small harbour and the Foryd Railway bridge, the sling net was a 200yd-long net fished from a boat on the ebb tide. Sometimes they ventured as far at the Point of Ayr. However, like the Usk tuck-net and other variously named nets, these are all drift-nets of some shape and size. We shall learn more about drift-nets when we come to the whammel-nets of the River Lune. But, for now, readers are probably feeling 'weiry' and so it is time to move on again to keep an appointment with a fellow who fishes the River Dee!

PART 3

NORTH-WEST ENGLAND

16

DRAFTS ON THE DEE

I remember some years ago watching fishermen in the harbour at Porthmadog running a seine-net around from a small rowing boat and, considering they were within the harbour itself, it seemed a pretty pointless labour. Unsurprisingly, they caught nothing, though they were obviously hoping for a migrating fish coming down the River Glaslyn. The same year I was poking about the River Dee at Chester after being asked to write an article about the Dee salmon boats. Back then it was difficult to find anyone to talk to about salmon fishing. But then, more recently, I came across Tony Randles and we met one Sunday afternoon alongside the river in Handbridge, just across the river from Chester, where most of the Chester fishermen had lived. Names such as Gerrard, Buckley, Price, Spencer, Totty, Johnson, White and Bellis joined Randles as names of families fishing by passing down the traditions through several generations.

Before arriving in Chester I'd read up a bit on the fishing. I knew the salmon netting had been bought out in 2008 by the Dee Fishery Association and that, by the following year, the fishery was now effectively closed, and remained so for the foreseeable future.

However, reading that in 2019, since the salmon season had closed, anglers have been reporting salmon ascending the rivers Dee and Conwy in large numbers, I thought it prudent to contact one – or some – of the traditional fishermen to see what was going on. I'd also read that poaching on the estuary of the Dee was rife because of the lack of enforcement of the catch and release policy.

Firstly, it appeared from the outset through various sources that a ban on netting wasn't the only answer to the lack of salmon. I quote from one source:

The absence of a net fishery on the Dee – with a declared catch in the 1990s of around 1,100 fish – does not mean that every year after the buyout will see increased returns at Chester or better catches on the rods. This will depend on the numbers of fish returning from the sea and a host of complex and generally poorly understood factors affecting abundance and survival in freshwater and marine environments. What the net buyout does mean is that a major source of loss of returning adult fish has been removed (the net fishery caught and killed around 17 per cent of all Dee adult salmon returning to the estuary in the 1990s).

As a consequence, irrespective of how many adult fish return from the sea, a significantly greater proportion should reach Chester and beyond because of closure of the net fishery. This is the lasting legacy of the buyout and it should mean that salmon (and sea trout) on the Dee are now better placed to face future pressures (including the potential threats associated with climate change) than they otherwise would have been.

Tony Randles is one of those dedicated fellows who has been fishing the Dee since before he can remember and comes from a family of fishermen. Four generations before that though, his family married into the Dobson family, who trace their fishing back many more generations. Not only was his grandad's brother, Jackie, an accomplished fisherman but so was his father, Paddy. Jackie holds the legal record for the largest fish caught, a 42lb salmon, although it has been said that a 74lb fish was poached from the King's Pool shortly after the First World War. Tony's two younger brothers, Kenny and Andy, also fish, while his son, Christopher, would probably have liked to if the licences hadn't been bought out.

We met as planned at the bottom of the cobbled part of Greenway Street, which today is lined mostly with new houses. One red brick terrace of three are original, the rest not. Many of the River Dee salmon fishermen lived here. When writing about the salmon boats, I can recall photographing some boats that were moored against the riverbank just below the Old Dee Bridge, and a few years later, when writing my book *Working the Welsh Coast*, the final words were actually jotted down while gazing out upon the river at this spot. Greenway Street is often referred to as 'The Lane' because its original name was Style Lane, named after William Styles, mayor of Chester in 1582–83. It seems that this community of fishermen dates back at least 300 years, even if the houses are very different.

Hauling in a Dee draft.

One of the main fishing stations called 'The Lane End' was right here at the end of Greenway Street, a bit below the Old Dee Bridge. There were many other drafts downstream and as far as Saltney Ferry, some 4 miles away by river.

'I was born by the lamp post up there,' Tony said, pointing up the hill. 'The council knocked them all down and built these. Those original three were privately owned so they stayed.'

The river was high after Storm Ciara and with Dennis just beginning to blow, it was probably going to get a lot higher. Britain was braced for flooding as climate change makes its mark and the government dithers after the realisation that the money put aside for flood defences is both not enough and hasn't, in the past, been used as it should. Add the building of houses on flood plains and you'll see that Brexit and the future fishing arrangement with the EU were the least of their current worries.

I looked around and saw how the riverbank had become overgrown. Established trees grew there and the area looked generally unkempt. I bet the rats were even bigger now!

'See over there,' said Tony, pointing to the trees I'd noticed. 'That was where we used to dry our nets. The Stakes, we called it. Each fisherman would have his own draft to hang them. Have you seen the photo of the area?' He quickly showed me a photo of the area with rows of nets hanging. 'Not a tree in sight. Nature has been allowed to overrun the area, which just shows how the fishermen would look after the riverbank, whereas now they've gone it just looks a mess.'

Tony showed me his nets. He had a trammel net for fishing down the estuary for flukes, and his draft-net, which he can't, of course, use these days. A draft-net consists of three stages: the higher part, the breast and bag, and the lower part. The mesh would be 4in, up to 6in in the wings. And he had an eel fyke-net in the van!

'Loads of eels in there but we can't fish them,' he added when he saw me looking. He pointed to his boats. One lies on the grass upside down, a small coble. The red boat moored in the river is his too. All the nets he makes up himself, adding that the trammel is his too and how his mesh is square when most use diamond-shaped mesh in the middle. Chester used to have its own cordwainer, a manufacturer of rope and cordage for the nets, but more recently this was obtained from dealers such as the Belfast Rope Company in Liverpool. Many of the clinker-built boats were built at Taylors Boatyard on the canal in Chester. I recall interviewing the last boatbuilder in the yard, David Jones, for a magazine article back in about 1996. He had taken over

the yard in late 1974 and reckoned he'd built a good few boats. A few others had been built by 'Old Boaty' in his small shed at Connah's Quay.

Three generations of Dee fishermen and three Dee salmon.

So I asked him to explain the fishing. He started off by saying:

Two men worked a draft, one man rowing the net out from where it is
stored on the aft platform (called 'The Shutes') of what was known locally
as a 'draft boat' (or Dee salmon boat), an open boat about 17ft 6in in length.
One man rows upstream a little, the 175yd-long net running out behind
into the river in a semicircle before bringing the draft rope back to shore.
Then a shout to the higher side man to walk down to where the lower side
man was. The call was 'I've got my gail', which refers to the end of the net.
So he'd walk down to the lower side man.

Sometimes a lad was given the task of rowing the boat, allowing the fisherman to control the shooting of the net. There was a 'master-cork' marker,
an empty one-gallon container, attached to the middle of the net over the
bag, and this was to indicate to the fishermen the centre of the net. As they
trudged along the grassy bank to bring both ends close together, it is said
that the occasional fellow got bitten by a rat, they were so prolific, their razor-
sharp incisors piercing all but the hardest of leathers. They didn't tend to
'walk' the net downstream in the way the Severn men do, perhaps for this
exact reason!

Then the net, 14ft in depth, was hauled in until any captured salmon was
trapped in the bag. They could always tell through vibrations and ripples
on the surface as the fish tried desperately to escape. Finally, with the net
pulled to the shore, and hopefully a salmon, the poor fish was flung onto the
riverbank and promptly dispatched.

As elsewhere, fishing traditionally started on 1 March and ended on
31 August. Fishing was allowed from midnight Sunday to midnight Thursday.
No fishing on a Friday, Saturday and Sunday, and often the first draft of the
week gained a good local audience on the riverbank. Even more popular
was the first of the new season that, if 1 March landed on a Friday, Saturday
or Sunday, this, too, had to wait until minutes after Sunday midnight before
fishing could commence. The time, then, to take on the salmon. This would
attract a huge crowd and, in times long gone, there would be singing among
the crowd. And he who caught the first salmon was awarded the Rector's
Prize by giving the local vicar the fish, a tradition that started in 1882, and
gained the fisherman a ton of coal. Later this was changed to one guinea.
Tony continued:

Handing over the first salmon of the season to the local vicar: the rector's prize being one guinea! The fisherman was Jackie Johnson, the vicar Rev W. Thomas with John James Evans, well-known Dee fisherman and Secretary of the Dee Netsmen Association, looking on, mid-1960s.

They tell us there's no increase in stock since we stopped fishing. So why did we have to stop if it makes no difference? Even when they cut us back when they introduced bylaws in the late 1970s, nothing improved. So what's the point? Still the drain off from the land is there, and fish still get caught out at sea. What we took makes no difference, but I guess it won't start up again.

Emotions I had heard many times before, and would many more to come.

There's a fish-counting station on Chester's famous weir, just upstream of the Old Dee Bridge, and a fish pass alongside, which gives its name to the riverside housing, the Salmon Leap Flats.

I asked him about the bylaws. Until these were brought in, there were eight drafts from the Old Dee Bridge (starting with The Lane End) and the railway bridge but the new regulations meant that the uppermost draft was that just below the railway bridge. They tried to ban all the drafts but the fishermen appealed and they were met halfway. This draft below the bridge was the draft 'The War Eagle', so-called from when, during the Second World War, some fellows were fishing there when a huge eagle landed on the bridge. When one declared that the bird looked like a war eagle straight from the battlefields of Europe, the name stuck. This was one of the deepest drafts at 12ft and usually avoided all the 'fassens', underwater obstructions that the net gets 'fastened' on. Sometimes they had to adjust their fishing draft to avoid these.

From the railway bridge to Saltney Ferry footbridge, there were nine more drafts and one below the bridge. So instantly almost half of their fishing drafts had been taken away, without any reduction in the cost of the licence. There were never any drafts above the weir. Tony reckoned that in the last year he fished it wasn't far off £500 for six weeks' fishing. The net would cost £200–250 and he caught two salmon that year!

Tony thinks the licences were first introduced around 1872. There were some forty-eight drafts and trammels for salmon. The rights were held by Welsh Water and their predecessors, and then the NRA. The licence holder had to fish and the area was from Chester to the Point of Ayre for the draft, and Connah's Quay to the Point of Ayre for trammelling. The Connah's Quay men had various drafts with names such as the Connah's Quay Hole, the Connah's Quay Lower and the Higher drafts but, as Tony says, the Chester men seldom ventured that far to draft-net.

Tony reckoned there was a good camaraderie among the fishermen. They'd repair their nets on the grass where we were standing. And often the fishermen would play cards waiting for the fishing to start, even in the dark hours of the night. Everyone knew everyone else!

It used to be the way that if you wanted to go out at midnight on a Sunday night, you'd go to the draft at midday and chuck your anchor out. It was a way of saying you were to be first but if you decided not to go, you'd say to someone else that you can have the draft. You'd choose your draft and others would wait their turn at a particular draft. You'd see a 'snoo' coming, which was the wake of a salmon, and your boat would be ready so you'd go out and shoot your net immediately.

Sometimes the fishermen would sail downstream and fish off Connah's Quay, or off the Bagillt Banks and off Greenfield. For this, the boats were rigged with a small spritsail. Some of the boats were based in Connah's Quay, and yet were of a similar build, the only slight difference being the shape of the transom, which was finer for the estuary boats to cope with the chop of the water. Mind you, this didn't seem to stop the Handbridge boats from sailing into these waters.

Fish would normally be sold to local shops or hawked around by the fishermen's wives, who were known to carry it in baskets upon their heads. The largest draft of salmon was in the early 1940s, when forty-nine fish were landed at The Lane End by Joe Johnson and his partner. All in all, it was a good week for them as they caught 308 in total.

As a last parting shot as we say our goodbyes, Tony mentions Trevor Jones, who still manages to fish his drafts on the River Conwy, by the road and rail bridges. I'd missed him on my travels, which was a shame, but decided not to backtrack. There'd be plenty of time for that soon.

17

THE RABBLE
FROM THE RIBBLE

You'd be forgiven for not expecting us to call in at Lytham St Annes on our journey north as the Ribble is a small (relatively) and innocuous river from a fishing point of view, although that fails to exclude the half dozen of fellows who, today, still make a living from venturing out, day by day, to bring home the harvest. In fact, so titchy is the fishing hereabout, it's hard to discover anything from written reports bar the occasional mention of salmon in the river. And shipping, of course. We'd passed through 'Shrimping Southport' on the way, and briefly observed the fishermen's compound with various tractors and not a lot else. I guessed the shrimping was a far cry from the time when fifteen horse and carts worked the beach, towing shank nets. Much of today's shrimping consists of pushing 6ft 'push nets' along the sand, parallel to the tide line, in shallow water.

Later, gazing out from the old Lytham lifeboat station across the estuary, I realised just how vast this estuary is – it's got to be over five miles across to Marshside, an area just north of Southport. And I know that area has some vestiges of fishing as the Lancashire nobbies – sometimes called Morecambe Bay prawners – were built at Marshside and, to add to our general knowledge of the area, there's a fine book by Harry Foster, *Don 'E Want Ony Shrimps*, the title of which sort of gives the game away! And, lo and behold inside he wrote, 'The estuary of the Ribble has long yielded a harvest of salmon for net fishermen.' One 85-year old, who died in 1972, was Hugh Baxter of Banks, who 'was licensed to set baulk nets' and other reports point to nets on stakes that would have caught various species including, in 1982, a 35lb cod.

Today's estuary is very different to the one that was constantly dredged for access to Preston Docks and, with the third largest tidal range in England, it's hardly surprising that the estuary, and river for that matter, have silted up once again.

To my left the river flows down from the hills below the bulk of Whernside, near the famous Ribblehead Viaduct, to which the river gave its name, to flow between the other two great Yorkshire mountains of Ingleborough and Pen-y-Ghent, some 75 miles upstream. Foster does give some incredible statistics for salmon fishing in the Ribble, where they'd fished 'from time immemorial' as far upstream as Preston. One was of some 15,000 salmon reported as being taken in 1867. A 1756 map of the river from 'Red Scar to Cliffton Marsh' shows thirteen fishing stakes in the river, although that doesn't necessarily mean they were just for salmon. Red Scar is well beyond the limits of what was the town of Preston at that time and today lies on the east side of the M6, on the north-east edge of today's city conurbation. 'Cliffton' (actually Clifton) is about 2 miles downstream from the entrance to Preston Docks. But so many stakes would substantiate the claim of thousands of salmon being landed on an annual basis.

The 1867 claim is slightly odd in that, since the 1865 Sea Fisheries Act had made fishing for salmon with a net illegal in the Ribble, this must have been either unlawful fishing (hard to disguise 15,000 salmon, mind), a wrong date or omitting from mentioning a delay period. But it just reinforces that salmon fishing was healthy at one time in the Ribble. Today the river is a major breeding ground for Atlantic salmon.

Reading further into fishing in the estuary, it appears that it was a haphazard set-up, a hopscotch of methods and people, all from the very lower class: a right rabble as somebody suggested. Mind you, if they managed to hide 15,000 salmon from the authorities, then they must have been made from strong stuff. Mackerel seemed to have been commonly caught in quantity and, at times, overwhelmingly so. But, on the whole, Southport fishing was about shrimps, the book title giving this away, and to a lesser degree cockles. Having said that, any fisher worth his mettle will find all sorts of fish in his nets if he tries. Stake-nets were commonplace at one time.

Foster's book also gives some insight into the detail of these stake-nets, of which we will learn more of in time. But for the moment, cast your mind back to the mud-horse fisherman Adrian in Bridgwater Bay and the stake-nets he set miles out in the sand. Recall me paddling in the mud, water pouring into my wellies!

So I stretch the ability of my eyes to focus on Southport, the roofs of which I can see from my stance: it would appear that the stretch of beach from today's Southport and southwards to Ainsdale was split into 'fishing stalls', some 1,044yd long and others shorter at about 360yd. Into these stalls their baulk nets were set up. George Masters, then aged 91, drew out a sketch of similar stake nets off Birkdale. Masters titles himself as 'Clerk of Works' and shows nets set in a crescent shape off Oxford Road where there was (still is) access to the beach and where there were 'railings for net drying'.

Back to the present, and back to Lytham's Old Lifeboat House Museum, which tells the story of how Lytham's fishermen first assisted stricken sailors before the first lifeboat arrived in 1851. Turn your eyes to the right and see the vast expanse of low-tide sands out to the point some 4 miles distant, and it's easy to see how people get into trouble on the sands. The nearby windmill, a familiar Lytham landmark, also houses a museum.

My experience of Lytham is about to get much better, thanks to Ian Lythgoe, who introduced me to the Ribble Cruising Club at their 'Dock' on the banks of a very muddy river that seems to be called 'Main Drain' but is a tributary that flows into the Ribble estuary close to the windmill. With boats of all shapes and sizes moored up into what certainly was deep, oozing mud, with berths accessible from a jumble of wooden walkways all higgledy-piggledy and boatyard facilities, this seemed to me a fantastic example of co-operation. And to top it off, above the boats and the moorings, was a grassy area, complete with barbecue and fine views over the marsh to Winter Hill and the moors above Bolton.

At Ian's insistence, various folk arrived from mid-afternoon to partake a beer and/or a burger from his barbecue. Thus, I was able to meet a host of folk over the afternoon and evening, chatting fish and boat talk. We weren't exactly a rabble, though some might disagree if they'd seen the empty cans of beer, but it was certainly a place to hear a good bit of what is now the remnants of a once-proud fishing hereabout: the Rubble of Ribble's fishing might be a better description! Much of the rabble might have given up the fishing but there were still a few who persevered.

Paul Sumner is one of those: he has fished the waters around here since he left school in 1977 and today is a bass, cockle and shrimp fisherman, though he admits shrimping plays a very small part of his year, much due to the lack of these in the bay. Bass fishing is the most profitable, which is why he described his Orkney fast-liner as his bass boat. It is shallow, outboard-powered and light to get over the banks where the bass are, he said. But I got him talking about the tasty crustaceans; shrimps, he says, are for low-water fishing:

A typical tractor rig for trawling for shrimps, this one at Southport. The two arms are lowered to drag a net either side.

Fish at high water and the shrimps are spread out over the estuary, so that's not a good time though there are some who fish then. At low water they congregate on the sands when there's no current so I go out a couple of hours before then and fish for two hours over the slack tide. You'd drag for thirty minutes and see what was in the net. Sometimes I'd have a drag back on the flood. At Morecambe they fish over the high-water and then get back to Morecambe before the tide ebbs but here we are exposed to westerlies and high water over the banks can get a bit nasty.

Once hauled in, then the shrimps go from the net into the basket before being riddled to separate the small ones that go back in to grow. However I've seen a sharp decline in the number of shrimps over the last few years, since about 1995 I guess, and don't know whether that's a natural cycle, climate change or over-fishing.

Paul started fishing for shrimps with an old tractor after school, then upgraded to a boat when he was 18 and three years later a bigger boat. He continued:

Shrimping is spring and autumn, say March to June and again September to Christmas. The summer is too hot for them and they go out into deeper water. When I started there were twelve boats shrimping, 25ft clinker boats, boiling the catch aboard. And three tractors as they can drag in 6in of water. They used to have twin-tailed nets – like trouser nets, if you will, so each can be emptied from the boat. Now it's single-tailed mostly.

Then, between seasons, some of the fishermen would turn to the salmon, so it worked out well. What about the salmon population I ask, for I'd heard that, if anything, young salmon were on the increase in the Ribble, albeit a very small increase off rock bottom:

I've never fished salmon but I've two mates who were stopped from netting last year. Up to then there were six licences for drift-netting, two in Lytham, two in Hesketh Bank and two from Freckleton, and that's been the case since I can remember. Don't know if they ever got any compensation. But the number of seals out in the bay must be having an effect on the salmon stocks. And the cormorants, and I bet between the two groups they take far more salmon than the six netsmen. But it's always the fishermen that get blamed.

Dave Morrison has been fishing for almost as long as he can remember. Today they call him Tractor Dave, though I never quite discovered why! His clash with Ann Widdecombe is worth mentioning. Back in 1985, when licences were being issued from Fleetwood, Dave was ill. The day after the deadline for applications, he found himself being denied a licence, even though he'd been fishing for years. So he was sent to London to fight his corner, ending up in an office near Whitehall. After a few refusals, he ended up in the office of Ms Widdecombe, though why she was there is unclear. Nevertheless, when he explained his predicament, she simply answered, 'You must be seen not to be making a profit.'

'I'll be happy to oblige,' was his answer, and thus he continued working the same way for thirty-five years.

He's worked a tractor for shrimps and a boat. He says it's a bit mad really, as having a licence for a boat and not for a tractor is nonsensical. You can tow three 10ft nets from the back of a tractor with no licence and yet only two 10ft nets from a licensed boat.

A basketful of shrimps, about 3–4st, are generally boiled aboard for some fifteen minutes. He tends to peel his own as he says the days of women gathering around in groups to peel them are long gone.

'They can earn good money doing other jobs, so why do the peeling?'

In a shell they will keep for about three days, and a bit longer once peeled, as the juice makes them go off. There's a chemical that can be added to prolong this period to around three weeks but I decided not to explore that further!

Today there are six licensed bass boats at Lytham and one shrimp boat, the latter belonging to Russell Wignall, who came down to the dock with his son Ryan. Russell, older than his looks that give the impression of a mixture of shyness and cheek, doesn't drive and is renowned for cycling to work on his old bike. He's also regarded as the town's last full-time shrimper and a sculpture in the town's Lowther Gardens is said to represent him, so much so that he had the honour of unveiling it in 2003.

But Russell also fished salmon, or at least had done until the ban in 2019. The year before there were only four salmon fishermen and they'd initially been given twenty-two pink tags for their £428 licence fee. A week later, each fisherman was visited by the EA, who asked for the tags back. Moments later, in what is now regarded as the biggest insult, they were each 'rewarded' with twelve red tags. Without any negotiation, and having taken their licence money, they simply reduced the number of salmon they could catch. Surely acting in such an underhand way is criminal? The following year they were refused any tags and their licences revoked, with a compensation of some £700 being offered. One fisherman from across the river acquiesced, simply saying he couldn't be bothered. But Russell refused, as did Andy Porter of Lytham and Harry Whiteside of Freckleton, the other two who today still refuse such a pittance in compensation.

I asked Russell about the salmon. They fish the ebb, which generally lasts some nine and a quarter hours, which gives them a window to fish five and a half hours after high water for some three and a quarter hours fishing. He uses his bigger boat, while others use small 13ft rowing boats. Fishing must be worked without a motor with a 150yd-long net that shouldn't enclose the whole river.

I also mentioned I had read about the haaf-net on the Ribble but no one, including Russell, had heard of this. One suggestion was that maybe, back in the 1800s, haaf-nets had been used on the river at Preston, between the two bridges where there's some shallow water, although someone piped up, 'You'd have to get off pretty quick when the tide turns as there's a bore!'

As the darkness fell, we searched the skies for the comet, its trail of light making it appear to be plummeting earthwards. The company thinned but the fire grew as we threw on old bits of rotten walkways and discarded logs. Then it was down to three and the conversation moved well away from fishing and the river. But it had been a wonderful evening and, with the tide high and the swirling water swishing around the hulls of the boats, the familiar clack-clack of the rigging, and the drifting remnants of smoke from the fire wafting through the open door of the van, I fell asleep.

Tommy Threlfall is an endorsee on Harry's licence and has been since he was 14 years old in the mid-1960s. His father was a shipwright in the local shipyard in Freckleton working on lifeboats and working vessels. There's a host of history of shipbuilding and all the ancillary trades that once existed on the banks of the small dribble of water called Freckleton Pool and Tommy must have been brought up among the last vestiges of it.

Today Tommy is a councillor on the borough council, and in the cabinet with the environment portfolio. As such he's able to gather attention to the plight of the salmon fishermen. We met the next morning and sat on his patio looking out onto a grassy field, with his cows preparing for oncoming rain, and it seemed so green and far away from the river. Nevertheless, salmon was upon his mind, despite the various phone calls about council business.

Andrew Porter netting on river opposite Lytham Windmill on a lovely summer's evening, having just shot the net at slack water. There was nothing better than going salmon fishing on an evening like that and watching a big fish go into the net. It was the best time ever for relaxing, something I don't think we'll ever see again; very sad indeed. (Andrew Porter)

The salmon fishing operates from Naze Mount, which they refer to as 'Four-and-a-half mile', downstream to the end of the river. This confused me when I first heard it from Russell, but it seems that Naze Mount is exactly that distance down from the centre of Preston. However, that run of river is the limit of the drift-netting, so Tommy's suggestion to Inshore Fisheries and Conservation Authorities (IFCA) that they (IFCA) take over control from the EA for this stretch while the EA retain control over the river upstream of Naze Mount makes perfect sense. Mind you, there are many among the fishermen that regard IFCA as hardly an improvement on the EA and reckon that, however you look at it, it's flogging a dead horse.

But to further this, Tommy has written to the local MP, who in turn has written to the current Environment Secretary George Eustice, mentioning the landmark case where the Supreme Court ordered the EA to pay compensation to Nigel Mott for the loss of his rights in regard to the putcher ranks on the Severn. Basically, he says give them back their fishing and tell the EA to stop their high and mighty stance or cough up a reasonable compensation, though the former is preferable. Tommy awaits a reply from Eustice. But for now we'll ignore the point that governments don't tend to err on the side of common sense, just rhetoric and lobbyism!

Tommy says that the attitude of the EA has been one of 'holier than thou' and that they simply alter the rules without any negotiation, constantly repeating their claim of upholding 'the interests of enhancing salmon'. The trouble is that all the various boards, trusts and agencies are run by men who are landowners or enjoy a spot of fly fishing. They simply have no understanding of working people and that folk catch salmon as part of their livelihoods. He says he remembers the licence being £60, then increased to £80 then £100. At the same time they used to start fishing in March, then that was taken away, then April, followed by May. Weekend fishing was banned from 6 a.m. Saturday to 6 a.m. Monday. The licence holder had to be in the boat and not just the endorsee. A constant altering of their working abilities. What other trade or industry would put up with this attitude?

In the same mouthful they harp on about how they have improved the river for salmon breeding. They used to say they were raking the gravel beds and making other river improvements such as fish passes, but even all that has now stopped. In fact, Tommy says they are doing absolutely nothing to enhance the salmon stocks. Salmon stocks, on the other hand, have vastly improved in the river, attested by the number of seals and cormorants feeding on them. But first the cheek of taking the twenty-two tags and replacing them with twelve, and then the total revoking of the licence simply breeds contempt for these officials and their insensitivity to working people.

Unfortunately, Andrew Porter was away during my visit. I caught up with him online and asked him if he always fished down from the Naze Mount.

No. Sometimes we fish down river or from Church Scar slip, near Lowther Gardens, where we moor our boats: that's me and Russell. We had to punch the tide to get to the Naze Mount and would probably take an hour and a half, depending on the size of the tide so sometimes we went the other way and went with the tide on the ebb seaward and then back in the flood. But usually more fish drifting from the Naze as it's narrower up there, but again would depend in the height of tide and weather was a factor.

So, you would start to shoot your net at one side of the river, net to the wall, then row across while shooting it. When the other end, starting to get ahead of where you are, you would drop the end you had hold of and row back to the other end and try and hold it in the tide while rowing and then let the end you've just let go of catch up, you row into the tide holding the net there as best you can. Sometimes it was best just to let it go for a while if there's a back to it, but it takes a bit if practice.

There's a small buoy on either end of the net with floats all along in-between and when we fished at night we'd put a flashing light on both ends.

To be honest, I think we've all been stitched up by the EA. Almost twenty-five years I had that salmon licence, which had been passed down from another old timer when he died. The thing is the rod fishermen upriver are still able to keep two fish each and the last I heard was there's almost 1,000 rod licences. That's 2,000 fish.

Just in this river, we have been left with nothing after a long tradition from way back. Some of the rivers north of here, including the Lune, have haaf-net fisheries but they can only take trout. I tried to argue the case and give Russell and me a haaf-net fishery as Harry Whiteside is too old now, but the EA were having none if it.

Another thing is poaching, as there's no bailiffs around here now. There might be one, I think, but I've never seen him for years. Yet poaching is going on right under our noses and people don't care because they know they won't get caught. It's all crazy. I heard the other week that there were two guys boasting about what they were catching and even showed Russell a photo of one catch. Talk about rubbing your nose in it. Now, even worse, we've had news that they are going to try to make the Ribble Estuary a Conservation Area. That will kill the fishing here altogether if that happens: no shrimp fishing, bass fishing or anything. The Environment Agency will have won.

And I'm sure it won't be the last time.

18

LUNE-ING ABOUT

It was June, bright and airy, and the salmon fishing season was under way on the River Lune. We – the kids and I – arrived at Sunderland Point in time to drive over the causeway just after the ebb tide had left the road. For Sunderland Point is at the very end point of the river, where it emerges into the wide expanse of Morecambe Bay, and is one of the few places in Britain that gets cut off by road from the rest of the country at most high tides. The signs are obvious, both those placed by the local council at the point where you start to drive over the rough marshland, and by the amount of grey thick sludge lying on the tarmacked narrow road. The tide can rush in at a great speed, up the muddy channels and onto the road and so all drivers have to be well informed as to the times that the road is passable.

I'd been fed that time via e-mail by Sandra, Tom Smith's wife. I'd known them a number of years and, back in 2009, when writing my book *Fishing Around Morecambe Bay*, had dedicated a chapter to Tom's life as a proper longshoreman, living both off the sea and river as well as the fruits of his labour in his market garden. His would be considered an idyllic life by many, although it certainly wasn't without its hardships and hard work. Tom had been born into a fishing family only several doors down from where he lived with Sandra and their son Thomas.

Tom's grandfather had moved over from Glasson Dock, just across the river, in 1901 and eked out a living by fishing and farming, and Tom's father followed on, as did Tom. Born in 1933, he started fishing with his father at the age of 15, and never stopped until he had a stroke.

Back then, we'd sat in the front room with a tape-recorder as he chatted about his days. Sadly the tape got lost but the transcript is the chapter in the book. But, as I write, I can clearly picture Tom out on the sands with his horse Prince, as he trawled the shallows for shrimps. There's a short film on

Vimeo entitled *Netting the Tide*, which follows him out on the sands. Prince was 33 years old when he died and Tom packed up the shrimps; he had had the horse for twenty-nine years. Now I recall how he described the way he trawled out there, walking 4 miles out into the bay with Prince hauling his cart, and then 4 miles back afterwards. Back then he told me:

Father never fished with a horse, didn't want the trouble. I bought a horse as soon as I could afford one and I was the last horse and cart fisherman here shrimping. I didn't do the cockles but it was very successful shrimping, from Heysham Harbour, what we call Heysham Lake, over the sands, going up towards Middleton, and then out. Just water, horse and weights and going in the dark, very often ankle deep. The finest shrimps ever seen in Morecambe Bay down there over the years. Absolutely superb they were. We used a 14ft trawl. I didn't use the shanks like they do over the Bay with the tractors. I use a straight trawl that could transfer to a small boat. Normal beam trawl, fastened to the back of the cart, on a bit of a run of rope so it could adjust itself and we trundled into the water. I had a super little horse, he was superb, led me a dance sometimes when we were away, not when he was running free on the marsh. But he'd tackle anything. And it was only an autumn job, so it was September right through to about Christmas, so we were going in the frost. I took him when it wasn't fit out and he always tackled it. But they weren't always there. There were bad seasons as well as good. I've seen ten minutes fill a trawl up. I once dropped into a gulley that ran out over the sands and it had created a little bay, and I thought it quite firm, there's been no great deluge of rain or anything, and I trawled across the rough pool, it's called, well up towards Heysham, and I said what's to do old lad you're full, and just that I had a fish box on board, like a flat cart, rather like the rag and bone collector would use, you know, in the old days, and I filled a 6st fish box. The net was behind the cart and the cart had all your gear on, fish hamper. You pulled to the side. Your horse stood just out of water and you tried to catch them before they came out of water and get the sand out of them. And then you'd have a fish hamper on board to tip them in and then walk back in the water to swill all the sand out of them. Empty through the cod end. Like with the boats. When I first started there was some lovely plaice and turbot down there, you could pick them up with a horse. They were so shallow. I mean you went in up to 2, 3 or 4ft sometimes, but they were sometimes so far inshore that you went in the daytime late in the season in November you couldn't find them with the boats, you couldn't get onto them, but you went in the evening tide with the horse and they were as thick as pitch, they were absolutely solid. They were there somewhere and you couldn't get close enough for them. I've seen it with the iron on the side, trawling on a little bank side, the one trawl iron, so a beam that high from the irons actually showing above the surface, and the shrimps were in that. You could feel around your boots, they were brilliant, my father had never seen such quality. You'd bring them home with a box lid on or a wet bag on, take them into the building with the light on because you brought them home to boil as against boiling aboard the boats. You were only away four hours so everything was fresh. And you'd take the lid off the box under the electric light and they'd be all over the floor, they'd be changing colour, taking on a paler hue under the light while they crawled away across the concrete. Like little monsters they were. They're beautiful, they were. I liked working nights, digging under a full moon. It's sad it's declined so rapidly.

But it wasn't just shrimps he fished. Whitebait and sprats at times, flounders, which he reckoned TV chef Rick Stein popularised and this helped his sales. Flatfish in the river, mussels from the skeers, and cockles in season. Much of the mussel catch was sent by train from Glasson Dock over to the east coast, places like Scarborough and Filey, where they were used to bait long-lines. But the summer salmon season was often the icing on the cake, until stocks depleted. He'd whammel-net (of which we will hear more in the next chapter) and haaf-net.

But we were here this time to learn about haafing, as they call it, rhyming with 'paving'. Over ten years on this was. The kids – Ana and Otis – had been up with me to Sunderland Point the previous autumn and they'd eaten hearty breakfasts in the same room that I'd taped Tom. Sadly he was upstairs, sitting in his chair watching the river go by after a stroke, though I'd been able to sit with him and chat away again about life and fish. I remember Sandra coming up to see how we were and I asked about what the kids were up to. 'Happily eating cake,' was her reply, which didn't surprise me!

This time we did the same, the kids eating bacon and egg and probably more of Sandra's delicious homemade cake, and me chatting to Tom. But, as I said, it was the haaf-net we'd come to watch in action, so it wasn't long before, with the tide towards the end of the ebb and our stomachs nicely replete, we were down on the foreshore, gingerly stepping through gloopy mud.

Phil Smith in action with his haaf-net at Sunderland Point.

Margaret and Trevor Owen have been haafing since about 1990, owning two of the twelve licences for the haaf on the river. Phil Smith has another, though for not as long. Margaret and Phil were dressed in waders and yellow oilskins, or 'yallers' as Margaret calls them. Trevor wasn't playing as the salmon were scarce and they were as much getting into the water for us to see and photograph than actually being convinced they'd catch anything.

I didn't learn how old Margaret was but her face was a picture of health and age, weather lines etched into the brown skin, and she's slim, though holding her 16ft haaf-net over her shoulder you'd think she might fall over. Into the river she waded and, waist deep, she lowered the net so that the central point dug into the seabed, unwinding the net as it went in.

'Salmon run upstream on the flood tide,' she had said, 'and turn back when they meet the cold flow of the river, so you stand in the run to catch them. When you see them, and they are in the net, you lift and twist to capture it.'

A big salmon makes a wake in the water, which you can usually spot, as I remember the lave-netters telling me. I guess there's a similarity in both methods, especially as they were both unusual portable methods of fishing, though I'd suggest the haaf is older. Tradition has it that it was brought into these west-facing rivers by the Vikings as 'haaf' is generally accepted as being a Norse word for 'deepsea', which somehow doesn't explain why these river and estuary fishermen adopted it. There are some suggestions that it also means 'channel' in Norse, though whether this is simply to suit the method is unclear. In Shetland, the 'haaf fishery' is the offshore fishery where small open boats – the sixareens – sailed out to set lines for white fish, sometimes 50 miles out in the wild North Atlantic, staying out for two or three nights as they slept in the bottom of these 30ft boats. But the Viking 'haaf' remains as a local lore and the fact that it's probably been in use for more than a thousand years from when the Vikings came and settled in these parts makes absolute sense. It is, as they say, a very basic set-up.

It consists of a stout wooden pine pole, about 16ft in length and 3in or so square that is rounded except for short lengths of about 6 to 10in at both ends and in the middle. Three holes at these latter points allow the passage of sticks to form a rectangle, two end sticks about 5ft long and one almost in the middle (it's actually just off centre to allow ease of carrying) that is nearer 6ft 6in long. These are made from Greenheart, which is both strong and heavy so that they stay down in the water. The middle stick protrudes the 18in up and acts as a handle. Of course, there are different designs so that in some the beam is rounded all the way. Some even are made from aluminium and fibreglass. A net is then tied onto this frame so that it forms two bags, one either side of the middle stick, when held in the flow of the water. Today these are made from monofilament netting, which is bought and constructed to form the bag.

So there we were, standing on the edge of the river, Margaret way out on one side, and Phil nearer as he was being filmed by Dave Nash. I was trying not to let the water overflow my welly boots, somewhat unsuccessfully I discovered when I removed them sometime later. The flood was by then pushing Phil upstream, the tide having turned, and he had to lean onto the middle stick to stay upright. As the tide rose, we had to move landward, such was the rush of incoming water. Margaret, far too experienced to get caught by the tide, was moving back across the river. There were no fish, they both were saying, and even if they had caught a salmon, it would have had to be released immediately as there had been a 'catch and release' restrictive by-law in place since the start of the 2019 season. Neither fisher was happy about this, as I was about to hear.

Back on land and walking back along the grass where a line of the net poles for drying the salmon nets stood, Margaret vented her fury at the legislators. It was the same argument I'd heard before. Using conservation of stocks as an excuse, the authorities were, little by little, attacking the traditional commercial salmon fishermen while the rod fishers and their sport fishing, or angling, were on the increase. One 2008 statistic put the commercial fellows having just a 13 per cent share of the fish compared with the anglers, and that was ten years old!

By the time we reached the lane down which Trevor and Margaret lived, we'd been treated to a very full explanation of why they both thought it downright unreasonable and totally unacceptable that they weren't able to catch salmon after spending over twenty-five years fishing. Sea trout they could, but they just weren't the same, she declared. And, of course, she was dead right. Phil Smith was nodding his agreement at every corner of the discussion.

The kids had, by this time, re-joined us after they'd got their feet all muddy, which meant a bit of a wash down. Then Dave suggested a dog walk – he had three to our one – and so we walked down the path between the trees towards the edge of Morecambe Bay. Work along the foreshore had progressed since our last visit and a new stone building had been built and recently opened to the public. This building features a camera obscura, which throws an image of the horizon into the inside white-washed walls. Created by Chris Drury, a landscape artist with an international reputation, the building is the work of master craftsman dry stone waller Andrew Mason, who turned reclaimed stone into a beautiful chamber, and inside an ever-changing, upside-down and divided circle of sea and sky is projected onto the lime-washed wall by the lens built into the sea-facing wall. This is part of Morecambe Bay Partnership's Headlands to Headspace arts commissions around the bay. We spent many minutes watching the images, with the kids playing by peering into the lens from the outside. Further along there's a new structure for watching wildlife out on the salt marsh, rich in its habitat of bird life.

Nearby, Sambo's grave has been discreetly walled in, although the grave continues to attract visitors from all over. Sambo, originally from the West Indies, came to Sunderland Point around 1736, the servant of a ship's captain, and died in Upsteps Cottage. Sunderland Point, at the time, was the busy port of Lancaster, where Sunderland's residents and merchants had built warehouses, a jetty, an anchor smithy, block maker's shop and a rope walk, all to accommodate the small trading ships and their cargoes of all number of imports from mostly across the Atlantic, including sugar along with contributions to the slave trade. The first bale of cotton was said to have been brought into Britain through Sunderland Point. Mahogany, along with other exotic timbers, also came in, providing the famous Lancaster furniture maker Robert Gillow with wood for his creations, which went far and wide. The company became Gillow & Co., before becoming Waring & Gillow in 1903.

Shipbuilding flourished in nearby Overton and, in its heyday, it is said that Sunderland Point surpassed Bristol in shipping, though, with the opening of Glasson Dock in 1787, the port business disappeared almost overnight. Today, however, Sunderland Point is home to some sixty residents and, walking around, it is fairly easy to spot the houses converted from the old warehouses that line the old quay. On the other hand, imagining that this place contributed to the dark history of slavery too, by taking such goods as furniture to West Africa, which was traded for human misery, who themselves were then carried over to the sugar plantations and sold to buy more timber, rum, tobacco and cotton, is sometimes hard. However, it should never be forgotten and joins the realms of Liverpool, Bristol and Glasgow in the trade. Perhaps looking after Sambo is some form of reconciliation.

Whether (as oral tradition has it) Sambo died of a broken heart because he thought his master had left him when he went to Lancaster, or he died of a sickness on arriving in a strange country, or maybe through some other cause, is today not known, but he was buried outside of consecrated ground as he had not been baptised (though I'm not sure how anyone knew that after his death and maybe it was an assumption). Anyway, he was buried and since then his grave has been well tended. We stopped to admire it, then continued our walk around the very tip of the Point, gazing out over a sparkling sea to the lighthouse on Plover Skeer where, nearby, there had once been a very efficient fish weir. The dogs chased each other and the occasional stone, and splashed about on the edge of the water that had risen considerably. The expanse of Morecambe Bay spread out in a vista before us and the wavelets of the estuary contrasted against the brown of the muddy water.

19

WHAMMELLING
THE LUNE

Whammelling – it sounds a bit of a punishing business! However, the roots of the word are tricky. I've found various other words or phrases to explain a whammel, such as 'confusion' or 'to turn over', but the best surely comes from canal cut talk: 'I'm taking the whammel up the cut' means I'm taking the dog for a walk along the nearby navigational waterway! I regret I never asked Tom.

Whammel-netting is a form of drift-netting (or hang-netting as it's often called) that seems to have originated in the rivers of Morecambe Bay and most likely in the River Lune. It is known that two Morecambe Bay fishermen by the name of Woodman and Willacy took whammel-netting to the Solway about 1855, which shows it was already in existence prior to that date. Tom Smith says it was basically a form of drift-netting as the fish become trapped by their gills.

A distinct type of boat was developed for whammelling, one of 17ft in length, very shallow in draught and flat in the floors, and rigged with one high-peaked boomed lugsail and a small foresail. That was before the days when engines were fitted.

The method, I remember Tom telling me, was worked three hours after high water when the whammel boats make their way down river to the 'baiting buoys' in the estuary. Once there, starting from one side of the channel, the fisherman makes his way across river, paying his net out as he goes. On reaching the far side, the sail is lowered and boat and net drift down on the ebb for about a mile. Beyond this there is a shallow bar across the riverbed and the net must be hauled in to avoid snagging. Once across, another shot will be made before hauling in and heading back up for another go.

On the spring tide, the first boat may get five shots before the tide is in flood, the last boat only three. On a neap, no boat will manage more than one or two before the tide turns.

The intention of whammelling was to meet the salmon coming upriver to spawn. To the layman it sounds easy – just drifting down while the fish swim into the net – but, like most of these traditional fishing methods, it was highly skilled work. The nets were very light, even though they were 300ft in length and 12ft deep. To hold them in shape, they were kept afloat along the top rope with floats and held down with small weights along the foot rope. Both boat and net must be controlled in a line across the channel as they drift down tide: if the net goes faster, it will collapse in the middle, if the reverse is allowed to happen it will stream uselessly behind the boat. For Tom, this meant keeping the oars shipped and dipping them to speed up or slow down. Toward low water, the net is allowed to curve to a 'U'. But as Tom explained: 'It doesn't matter if your net ends are in the slack because for a while the fish come down to meet the flood – if you've got a 'U' shape, they'll swim down into the narrow.'

Tom Smith at the helm of his whammel boat Mary, LR53, built in 1927 and named after his mother. The simple rig made the boat easy to sail, though more often than not he propelled her by rowing or sculling over the transom. His brother Philip sits in the boat. (Alan Smith)

I remember back in the days of *The Boatman* magazine when I used to write for them and Jenny Bennett went to interview Harold Gardner, who started whammelling with his father Thomas (then the pilot for Glasson) at the age of 14 when it was a profitable business and one boat would earn enough to support two men. It was until recently a method of fishing that required co-operation between boats – an aspect that has never been so strong as it was in Harold's early days. The fishermen of Glasson and Sunderland Point had formed a co-operative and were working in company. It gave them the edge over the upstream boats from Overton as a 'company' boat would always be first at the fishing grounds and would thereby gain the longest possible fishing time. The 14-year-old Harold was not, however, welcomed with open arms by the older men. Regarding him to be too young, the company drift-men refused to grant him a full share of the income and his father decided that they would fish on their own. Jenny, many years ago, kindly allowed me to use some of her material, though much of what she wrote I recall Tom telling me. Accordingly, she quoted Harold back in about 2001:

> We were outside the 'company' for two seasons before they asked us to join them (we were a bit of a nuisance I should think, being keen and always out), and they offered me a full share, so that was that. We fished in company for twenty years. Then more boats came along (there'd been five), and there weren't enough fish to support two men in each boat. Next the engines started and the rest is history.

But whammelling has all gone now, another method dumped into the history books. Amazingly, in 1948, the number of licensed Lune whammel boats was unrestricted but with incomes to be had from musselling, cockling, shrimping and other types of salmon fishing, there were only five or six working whammel boats. Then, that year, the whammellers had an excellent season, catching more salmon than had been known in living memory. The following year, encouraged by the 'bumper crop', there were eleven boats in the fleet and by the 1950 season there were thirteen. The controlling body, first the Lune, Wyre and Keer River Authority and then the Lune Board of Conservators, decided to restrict the number to twelve. But some of the fishermen thought that was too many and reckoned six was enough, otherwise there wasn't enough fishing time between tides to make a decent living.

I quote from Tom's recording from 2009:

Tom standing alongside Mary in about 2005. The boat has recently been sold and is being restored locally.

The licence is £400 a year now and they've put on another £100 odd over the last few years. Over the same period we've had some of the most contrasting years in my lifetime, 2003 you couldn't work in an open boat in daytime. The heat was tremendous for me and the garden went to waste. And 2004 it went the other way and 2005 was cold. And again in 2007 and 2008. Never known such a cold July as this last one, it was nearly freezing. I think August was the only month in the year when there wasn't frost in the UK. There was probably frost in Scotland on the high ground. Then the flood water afterwards. This year we had the heat to start with and now the rain. I used *Sirius*, if it was convenient. But, so often, especially when you've had a massive amount of floodwater, the channel is so shallow, it's taken so much sand off and put it in the bottom, that I have a 12ft punt I borrow from a friend, one of Professor Bill Bayliss's, a really handy, very buoyant, boat. Two of us and 300yd of netting, it's like a cork, it's super, and of course from the point of sculling across it's quite lightweight. So I've used that quite a lot, even to the bottom bar. About an hour and a half to row, with tide with you. And make sure you get back before it starts kicking up a stink. 2002 was some of the best fishing I've ever had in my life for salmon. I know I broke my own landing record and so it was a bit of a let-down to have the weather turn against us for several years. And normally I should have made it up with sprat fishing, whitebait. There's more shrimp boats rigged up here for part time than there's been for ten years and they couldn't get a trip in.

Up to its demise, thanks to the EA who took over from the National Rivers Authority (NRA), there were ten licences of which only a few, such as Tom, were making a living from it. Of those ten, eight were running with engines while the *Ivanhoe* and the *Sirius* continue to sail. In Tom's view there was little advantage in having a motor:

> You were nearly always going with the tide: out on the ebb and home on the flood and, while you're fishing, it's drifting. I suppose fishing under sail is primitive. I know it was tiring if the wind dies and you have to row out or back but I don't understand engines! If an engine packs up I haven't the foggiest, but if the wind drops, now that I do understand.

Despite this adapting to change, the basic method of whammelling never really changed. The season ran from the beginning of April to the end of

August with no fishing between 6 a.m. on Saturday and 6 a.m. on Monday, the restrictions being designed to protect the fish stocks. I remember Tom talking about the good and bad sides of restrictions:

> The weekend ban was a good idea. You can fish both tides for five days and that's enough. But they should change the fishing season. There is a natural evolution with the salmon that means they run ever later in the year. Now, the best of the fish are running in mid-October. When my father started in 1919, the best fishing was always in the last six weeks of the season. Between then and now there was a spate of spring fish and June became the best month. Now the spring is very quiet and in the autumn, just as it gets really good, we have to stop! They could have the whole of April for just two weeks of September but, no.

To emphasise this, he explained about water salinity. 'Another big problem for some years has been the very high rainfall. The massive amounts of fresh water running into the Lune end up here, diluting the saltiness of the estuary. The salmon prefer a gradual transition from salt to fresh water and many will loiter off shore until the salinity increases. But by then, the season may have ended,' he told me.

One time, with the kids, we were standing by the net poles. This reminded me of a photo of fishermen standing in their gear in front of a flax net hanging down. The caption had been written by Tom's brother, Alan, and he had noted how the fishermen would put on their oilskins and sou'westers for posed shots when a cameraman was around, and always stand on the leeward side of the net to keep out of the wind. The photo in question was taken in summer and Tom and Alan's dad stood on the left side and their grandad on the right side, which was unusual as he normally went off in the opposite direction when there was a camera about. The net drying was an old flax net, which you could see from the heaviness of it. Totally unlike the lightweight nets Tom used.

On a Saturday Alan recalled cleaning the nets of the seaweed and shaking out the sand. The nets would be left to dry until put back into the whammel boat on the Monday. Tom had two whammel boats, *Mary*, LR53, built in 1927 and named after his mother, and *Sirius*, LR33, built in 1923. Both came from the Overton yard of Jack Woodhouse, whose yard was in the middle of the village and where a variety of boats were built – whammel boats, prawners, mussel boats and pleasure craft. *Mary* was used as a tripping boat when Sunderland Point became known as 'Little Brighton on the Lune' and Tom's father would charge 2s each to take ten or fifteen people out and round the lighthouse, upstream to Glasson Dock and back, which was good money considering most folk around there were earning £1 a week.

20

FISHING THE OTHER SIDE OF THE RIVER

A year later, with both Tom Smith and Dave Nash having passed away, I was back on the river, this time at Glasson Dock, which, in effect, means a back track. I'd gazed across so many times, sometimes even from Tom's window, that it seemed appropriate not to forget this compact village built as a later port for Lancaster. Time to see the Glasson Dock perspective, you might say, as perspectives can be very different when a body of water separates two communities. OK, so we are technically retracing our steps, but only by a distance of less than a mile by the way the crow flies (or salmon swims?). Yes, by road it's a motor back to Lancaster and across the river, and along, but sometimes things are worth it.

It all started with a pretty innocuous Facebook post that, not for the first time, turned into a wealth of information. It started by me 'talking' (that's Facebook for posting) to Stephen Worthington and he sent me off in various other directions; not literally mind, but towards others who had fished.

But first his own bit of history. Dick, Stephen's father, and also his grandfather, Jack, used to do the haaf-netting. Dick, he told me, started in the 1960s but he gave up his licence due to ill health. He also did the whammel from 1994 until 2004, at which time he went back to haaf-netting until 2017. He passed away in 2018.

Stephen tells me about his dad's favourite haunts. 'He was very keen, used to fish the ebb opposite side to Glasson dock, called Wilky Odd, down to the Lighthouse for the first of tide, then finish off at Penny Hill, which is at the back of Ashton Hall,' which I later found to be by Waterloo Cottage on the Ordnance Survey map, a bit further up the river from Glasson Dock and along the old railway line that is now a cycle path.

A lovely peaceful scene of calm and reflection at Glasson Dock c. 1900. Two whammel boats lie at their moorings, as do two small Morecambe Bay nobbies, all counteracted by the schooner being towed in by the steam tug.

When I mentioned Sunderland Point and the fishermen over there he did add a story that I hesitated to mention but deemed to include as it demonstrates just how dangerous fishing with a haaf-net can be:

> My dad was fishing down by the lighthouse when he spotted someone in the water. Turned out it was Margaret Owen. She had fallen down a hole at Sunderland Point, and drifted downstream with her haaf-net beam towards the lighthouse. Luckily Dad was there and he got his beam and hooked onto hers and pulled her in. That was when she first started around 1994.

Funnily enough, a few weeks later I listened to a recording by Margaret, talking about this near drowning. I think she really thought she'd had it as she floated off down the river and out to sea. She says she stepped back to where she'd been before, but adverse weather had changed the seabed, so down she went. The guy fishing with her had gone off to get the coastguard and, as she floated, she realised the coastguard would phone their house to inform them. That wouldn't be much use! Still, she was so thankful that Dick waded out, even though he couldn't swim. Though, of course, she hadn't mentioned it that day a year ago!

When I asked him about other fishermen he added a few names:

Derek Aldren had haaf-nets at Snatchems near the Golden Ball pub, just below Lancaster Quay on the opposite side, which he fished on the ebb. Willie Armer at Glasson Dock will be one of longest now, and his dad haaf-netted as well. Derek Milner has a beam at Glasson Dock. Bernard Black packed it in a couple of years ago though he'd been haafing for twenty years. Bill Bailey fished near the station house.

Another contact, Val Simpkin, mentioned that she'd often seen some haaf beams at Nans Buck Cottage on the Lancaster to Glasson cycleway and sometimes by Waterloo Cottage just further towards Glasson. It turned out that Val had spoken to Stephen's dad on several occasions while he was fishing by Waterloo Cottage!

Thanks to the magic of Facebook, I quickly found Bernard Black. He told me that he fished the flood and ebb tides at Ashton Hall, where he fished off the shore there and also from the back of what was the now-defunct Victoria Hotel at Glasson Dock, and also further downstream of Glasson at Crook Farm, again both on the ebb and flood tides.

But Bernard also told me about draw-netting, which seemed much the same as the Chester draft-net or the River Severn long-net. He described it thus:

A draw-net is so many metres long, I forget the correct length. One man stands on the shore with a big end stick and the other rows the net across the river depending on ebb or flood and rows down the river until the net has all gone, then he has a long rope and rows ashore, when he reaches shore he starts drawing the net in, closing the loop, then starts drawing the net in by the bottom cord at the same time the other man walks towards him, keeping the net closed all the time. When they are about 20ft apart the other man draws the net in the same as the rower of the boat until all the net is pulled ashore, trapping the salmon in the loop and pulling the catch ashore.

It seems the draw-net was mainly used between Overton and Glasson Dock, but also Sunderland Point, back to shore at Crook Farm, and at other times a bit upstream between Sunderland Point and Glasson Dock. Bernard said that he'd fished this for about ten years before packing it up, while he fished the whammel for three years.

'Whammelled from Bate Haven Buoy near Cockersand Abbey to the mouth of the River Lune,' he told me, which was about what Tom had told me. Bernard remarked that he thought Tom always fished well, in the company of anywhere between seven and ten other boats, though he didn't know the exact number.

When I asked him which was his favourite way of fishing, he quickly replied: 'Draw-netting, because you had someone to talk to and it was good exercise!' Mind you, all that wading around in water with a haaf-net I'd have thought was equally good exercise.

On the Google map of the area I found a mention of Ice House Hill on the Ashton Hall estate. Researching further, I found mention of a Grade II listed ice house just to the north-west of Ashton Hall itself, on the edge of what is now a golf course, and very close to the lake, the western end of which is also close to Nans Buck Cottage. Presumably ice was taken from the river and, due to its proximity, stored in the ice house. According to Historic England, the ice house is probably nineteenth century, and is constructed from sandstone and stone rubble and brick, and is mostly covered by earth and vegetation. It comprises a domed area and a vaulted entrance passage, although the doorway was bricked up and the interior inaccessible at time of survey in 1968. Ice House Hill is shown as the slope to the south of the ice house.

Willie Armer, born and bred in Glasson, fished with Dick Worthington and so I found myself one sunny morning sitting outside the shop-cum-cafe by the lock chatting to him. Ironically the building and the cafe on the corner belonged to Bernard Black. Bernard had been busy the evening before as Ana and I partook of hospitality at the Dalton Arms and I was too busy with Willie to have another chance to meet him.

'He can be a bit grumpy,' says Willie and proceeded to tell me how, when he was running his cafe himself, he'd lock the door and tell anyone eyeing up being a prospective customer, perhaps even one about to open said door, that he was now closed and was off fishing!

Willie couldn't have been more helpful – something I've come across on many occasions during this voyage into west coast fishing and I'm equally told that, together with John Smith (Tom's younger brother), the two of them are a double act and they'd keep Tom and Sandra in stitches. He's 'a good chap' Sandra texted me!

He started fishing after school in 1967, a time when there were forty-six haaf licences on the Lune. Today there are five on this side: Michael Price from the smokehouse, Terry Franklin, Willie and two fellows from Lancaster. On the other side it was the Smiths, Braids and Gardners that fished in the main. Des Worthington biked all the way from Fleetwood to fish his haaf, he being Dick's cousin. Other names mentioned were Bill Bailey and Jack Butler, who lived in Morecambe at the time, while his nephew Tim Butler was still around. The Braids, he added, were the best people at the draw-net. My list of fishers' names was expanding!

Willie and I had a long discussion about the fishing stations for the haafs, called locally the 'odds', which were strategically placed along both sides of the river. They had names such as Nan's Buck and Penny Hill that I'd come across but then there were Tommy Odd, Wilkie Odd, Stoneheap Odd, Sevenstones Odd. To add to my confusion, Corner Odd was so-called on the Glasson side while on the Morecambe side (as he referred to it as), the Overton lads called it Kyle Odd. Some were fished on the ebb and others on the flood, while a few, such as Nan's Buck, could be fished on both. The oddest Odd was Katie's Breast Odd, near Fishnet Point. There's a good tale about Reg Sculps, a London lad who made good with a local lass from the Robinson family of fishermen who happened to be called Katie. Imagine how Reg would feel when he heard a haaf-netter declaring he was 'going to have an hour on Katie's Breast'!

Willie never calls himself a full-time fisherman but, like many others, is keen to keep the traditions of the river alive. Generally he's a surveyor, which enables him to fish when he wants to. He'd been out just two days before and had caught a 14lb salmon, which had to go straight back in. This led on to the subject of 'catch and release' and much of that conversation was ground already covered. He feels just as angry at the behaviour of the EA as those already spoken to. One point was the mortality rate of salmon caught by the anglers:

They don't realise but so often they've worked a fish on a rod and tired it, landed it and removed the hook before throwing it back in. Well, that fish might have suffered that once or twice before, and its chances of survival after the third time are slim. With a haaf, that fish is netted and then turned back into the river in seconds. Even in the unlikely event of it being landed in another haaf-net, it is still unharmed. But they can't see the difference.

We talked about the salmon themselves, and how the water in the river affects them:

> You can tell a lot about a salmon by just looking. A grilse – a salmon back in his river for the first time to breed – has a forked tail and gets called a 'forkie', while older salmon have a flat tail. Upper river salmon change colour and can get ugly and are sometimes brought down the river with a 'fresh out' due to high rainfall on the fells. These we call 'droppers' as they've been pushed down and get flushed out with cleaner water.

Much of the salmon caught in the days of old went to Morecambe, where strawberry and salmon teas were almost as popular as shrimp teas. And something I never really thought about was that grilse, because of their size, developed a stronger market, as Willie explained:

> Salmon used to get hawked around the area; grilse mainly. If you buy a big fish you are hard pushed to eat it. Too big, too much. OK for hotels but not for a household. Get a smaller one and you had to steak it up and keep it, especially before refrigeration when hotels might have a fridge but homes generally didn't. So grilse were always easier to sell.

I asked him what he was looking for when out standing in the river. Did he always stand in the same spot?

> No, not necessarily, you could be standing where that fence is [he pointed to some railings about 20yd away] and not get anything and think, just a minute, I'll move back, just out of the run of water. Things might look better than they are, with rain water it alters the sand, it narrows it down and puts the run into one place and the fish might follow that. Little things really. At night time the fish will follow the sand, rather than the stream, so certain places with sandbars might push the channel across and down one side, so you get to work at the back of it as centrifugal force pushes the fish across the river, both on the ebb and flood. At Penny Hill, we sometimes do better if the sandbank on the other side is higher to push harder towards

Penny Hill. I think the fish are using the water to save their energy, aren't they? Fish on the flood and the fish is going up but on the ebb that fish might have been pushed back when you catch it, or it's swimming down. What happens on the big tides, they might be able to get over the first weir [at Lancaster] and drop back. They go in stages, go up maybe 2 miles and wait, drop back half a mile. Remember they are changing from salt water to fresh water, so they have to acclimatise. That's often when we catch them when they drop back. It's just a matter of finding how the river flows. Somebody could be fishing there and with the tide over there, someone else pushing back behind them. End of the day it's luck. If there's plenty of fish in the river, you'll catch one.

There were fishing stakes in the river, just as there were in the Ribble, and probably the other rivers around here. Willie said the posts were still visible off Penny Hill, while the ones downstream at Bank Farm caravan park, a bit past the lighthouse, had all the posts cut off some years ago by jet-ski enthusiasts. Then I found a mention of stakes in the *Lancaster Gazette* of 1838. This published a letter from 'An Angler', who voiced his concern over the decline of salmon numbers in the Lune, and described some of the causes. One of these was an array of stake-nets along a mile of the mouth of the river, which he deemed was so numerous that the writer wondered 'how any fish is enabled to pass them'. The Lancaster Quay Commissioners had cleared the nets 'frequently', it said, but they were soon replaced.

We talked about these and the other nearby rivers. Willie again:

Don't know about baulks [as they call them] but when I was working on the motorway, setting it out between Carnforth and Killington about 1970, we bridged over the River Keer and there was a guy on the motorway there, every night taking at least two grilse, tickling them in the river. There must have been fish in all these small rivers.

On our journey north from the Ribble, seemingly a lifetime away, we had passed though Fleetwood and over the River Wyre just by Skippool (after a poke about Skippool Creek) and I had wondered about salmon fishing in the river. Asking around, no information came to hand, until Willie was recalling the different men who, over the years, had had licences on the Lune. He mentioned the owner of a garage called Storey's Garage, near the Shard

Bridge over the Wyre, who had fished that river with a haaf. When the river authority took away his licence to fish there, they gave him a licence to fish the Lune in compensation. So, yes, there were the odd haaf-nets on the Wyre, though Willie was unsure of the date the licences were taken away from that river. Pre-1965 we guessed. This led my thoughts back to the Ribble and an assumption that there probably had been haaf-nets on that river back in the nineteenth century or earlier.

Willie had found a brass licence disc dated from 1888 in the mud of the river several years ago. He showed it to me and it was stamped thus: 'L, W, K & CD', though the 'CD' could have been 'OD'. We assumed rivers Lune, Wyre, Kent and possibly Conder. This disc would have been fixed to the haaf, and pre-dated the plastic versions of later years, which then turned into credit card-sized licences of today.

I could have sat for hours listening to him, such was his enthusiasm and knowledge, until he suggested a drive around. I jumped at the chance and we spent another hour in which I discovered how the village used to be further

Willie Armer's brass lave-net licence tag dating from 1888.

inland, where the baulk or fish weir was by the lighthouse and some of the old tracks over the low land. When I mentioned the ice house at Ashton Park, off we were to find another on the nearby Thurnham Estate where the Dalton family used to live. It was they that had once owned the land on which Glasson Dock is situated and had sold it to the Lancaster Port Commissioners, who started construction of the dock in 1782.

So we wandered through the woods alongside the Catholic church, pretty mushy under foot, until coming across two small lakes, side by side, one having a very full covering of green duckweed. We spent five minutes hunting around until we found the ice house lurking under a mass of brambles and an uprooted tree. Fighting the brambles, and the odd huge stinging nettle, I found the entrance. Possibly 8ft wide, it has a vaulted roof and, although the end facing the lake was accessible by a full-height doorway, the rear wall was partially underground, with only a couple of feet between ground and roof. The tree had caused obvious damage to the roof and there were gaping gaps between it and back wall. Someone had built an inner wall with a window looking into the ice compartment, which made me wonder why. Still, it was obviously used back in the day, filled with ice from the two lakes, to keep salmon, and probably other food stuffs, cool.

Back in Glasson, we managed a quick pint in the Dalton Arms before parting on the understanding that Willie would draw up a map of the river with as many of the 'odds' on it as he could find. But we were finished here and the following morning, after a night of listening to the rain lash against the roof of the van, it was out with the bikes with a thin sun overhead, and a cycle up to the maritime museum in Lancaster, which I wanted to show Ana. It was closed, presumably as the council couldn't be bothered to reopen after the lockdown, even though the rules governing museums said they could reopen. But that's council-run anything for you; an attitude of 'we don't really care', even though we were well into the cut-short tourist season.

Still, we cycled back to stop off at Nan's Buck cottage, where I spotted a haaf-net belonging to Michael Price, partly hidden. Both Willie and Michael Price have another net each at the Back o'Vic Odd – also known as The Steps Odd – ready to go. At the end of the ride, we sat by the river for a moment, watching the water, the wind having risen bringing in the threat of more rain. Wavelets dashed themselves before us as the tide reached its high point. To the right the salt marsh of Conder Green, with its crop of samphire, had shrunk and across I could see the causeway to Sunderland Point well under water. It's hard to understand that here the river was probably brimming with 14lb salmon and yet no one was able to fish it. Sea trout in July would be scarce and none of the fishermen were going to be out today. Yet it's a wonderful little port, active with a seemingly constant trail of wagons coming and going into the dock, loading with animal feedstuffs and fertiliser, yet just as popular with tourists who love the calm atmosphere akin to a Cornish village. I promised to return.

21

HEDGE BAULKS AND GARTHS

I've already mentioned the fish weir at Plover Skeer, just by the lighthouse at the mouth of the River Lune. This was of the type with stone walls and interwoven willow on stout oak posts, called locally a 'hedge baulk'.

It is believed that these hedge baulks were extremely old and it has been suggested that the monks of Furness Abbey owned the rights. Before that, further suggestions have been made that it was the Celts who first devised them, before the Romans came and encouraged their use. More recently, ownership has been traced back to the early eighteenth century, while at least one baulk was owned by a family with no other connection to fishing. Presumably this was simply rented out to fishermen. These baulks have been sited for years all along the coast, from Heysham to the shore off Bare (the north-east part of Morecambe), and they consisted of an elaborate construction of stone, posts and hedging, involving a great deal of skilful labour and expense. One such structure was sited on the largest skeer off Poulton which consisted of five hedge baulks next to each other in a zig-zag way. The names were often localised according to past fishermen, such as Jacky John Skeer, and again one of the baulks on Old Skeer was called Old Dick Bond's. Even the Ordnance Survey map shows a few of them, including the largest, Aldren's Hedge Baulk, so named because it was last fished by Jack Aldren even if the rights belonged to the Pennington family. Another, alongside this, is the Old Skeer. Others are simply labelled 'fishing baulks' by the Ordnance Survey.

From Glasson we drove up to Morecambe and, with the tide low, managed to see the expanse of beach and where the remains of the stumps that were Aldren's Hedge Baulk lie among the sand. It was a sloshy walk out and, to be honest, not terribly exciting. Stumps in the sand are stumps. Full stop.

Out of the sands, emptying the Plover Scar weir.

To build a hedge baulk, the basic shape resembles a large 'V', with a wide-open mouth, which is laid out on a shelving beach. Stakes of oak – up to 12ft in length – are driven into the sand using a heavy wooden mallet so that some 8ft of oak sticks out above the sand. It is constructed at right-angles to the shore, and the length of arms of the 'V' are different, one being some 340yd long while the other is some 270yd and the mouth of the baulk is 150yd across. Between the oak posts is wattle hedging of hazel, against which the sand will eventually silt up on the outside. At the apex of the baulk is an intricate cage made out of netting over the posts, including over the top, where the fish end up. This netting has a stone foundation in which there is a grating to let fish out. Around this cage there is another low hedge along which water will flow out of the main baulk. There's also a sprat pool, where small fish collect, and a 'puzzle garden' which, to me, is indeed puzzling. Of all the various fish weirs and structures around Britain, these surely are the most complicated and thorough in design. They were, however, deadly in their catching of fish and 60,000 herrings in one go was not uncommon.

Morecambe Bay also has an assortment of other stake-nets. Firstly, there's another type of baulk net found all over the bay from Glasson to Barrow, as well as into the Duddon Estuary to the north. Again, this is a series of posts with nets hanging off but an ingenious smaller baulk of timber – a much thinner piece of wood than the post – is tied to the top and bottom of the net and to the post at the top. When the tide floods in, the baulk lifts the bottom of the net to allow the fish to swim in, while on the ebb the baulk sinks and closes the net. Other forms of 'fixed engines', as these structures are called, are the paddle-net, roa-net, poke-net, teedle-net and bag-net, all of which are similar in that they consist of suspended netting, though their setting and method of action are very different.

These structures trapped all manner of fish but especially herring, codling, plaice, flounder and whitebait. Herring was plentiful and were available at a penny for as many as you could carry. Herring was also caught in drift-nets set from a boat in the autumn and early winter and fishermen sailed as far north as Maryport. Salmon isn't mentioned much but presumably one structure such as at Plover Skeer, near the mouth of the river, trapped salmon.

We now hop north again, up the Cumbrian coast to Ravenglass, a delightful small coastal village perched on the estuary of three rivers – the Esk, Mite and Irt. The village consists of little more than a street lined with houses and two railway stations: one for the main line route up and down the Cumbrian coast and the other for the miniature Ravenglass & Eskdale Railway, which it's likely that the majority of visitors come for. But it was the Romans who first built anything here, finding it perfect as a port to supply their forts inland. They built their port in the second century, and for 300 years this was an important naval base, bringing in ships. Goods could be sent up the road over Hardknott to the fort there, and further on to Ambleside.

I'd purposely come to see the fish garth, the remains of which are in the river. The tide was low so the kids and I walked along the muddy and stony foreshore, down to the more solid sand to where the posts still stand up, betraying much of the shape of the structure. There was no need to linger and we took the route back under the railway bridge and past the old Roman baths at Walls to eventually end up at the car park once again.

According to the dictionary, a garth is an enclosed quadrangle or yard, especially one surrounded by a cloister (Middle English) and in this case as a small yard or enclosure and perhaps assimilating to the fact that the fishery would belong to the monastery. Indeed, one in the River Irt was referred to as the Monk Garth. Generally, in its early days, it provided fish for two abbeys and a priory.

One of the Morecambe Bay fish weirs.

Historically there have been many fish traps or garths in the Ravenglass Estuary and tidal stretches of its feeder rivers. The garth we viewed, generally called the Ravenglass Salmon Garth, and others further up the Esk, were established during the reign of King John. Permission to operate the Salmon Garth was granted to Muncaster Estate and for that further upstream to the Vicar of Bootle.

Garths, like some of the other traps we've seen, operated by allowing fish to pass upstream with the incoming tide but partially restricting their route back downstream on the ebb tide. Fish then became trapped in a channel and pools behind the garth and were caught by the operators using nets. This garth was, as most were, a rather complicated structure and the details of its operation have been lost. This one, last operated in the 1970s, was a simpler 'V' shape with its apex pointing downstream and wings going out to the sandbank on the inland side of the river channel and to the shore near the railway bridge mentioned.

In its latter days, the garth operated under a licence issued latterly by the Cumberland Rivers Authority. Fishing was effectively allowed only between Monday and Friday. A gate at the apex of the 'V' could be opened or closed to allow fish to pass through or not. The gate shouldn't be closed until 12 noon on Monday, and by 6 a.m. on Saturday it had to be opened. If the tide time was right, a buoy with a long rope could be attached to the gate so the drag of the tide pulled it closed. It's interesting that it was noon on the Monday and not 6 a.m. as per other salmon fisheries.

No metal was allowed to be used in the garth construction and wood from Muncaster Estate was used to repair it. Its walls were about 4ft high

Detail of the sluice on the Ravenglass weir, which can be opened to allow fish to escape during no fishing times.

and constructed of hurdles of 'reches' woven between posts driven into the riverbed, sometimes with nets above the hurdles in later years. Two-inch gaps in the hurdles were required to allow small fish to escape – although fish tended not to be keen on swimming with the pull of ebbing water and ended up trapped anyway, just the same way as the baulks were constructed.

Operating the garth was hard, physical work. Tides ebb and flow at times that advance every day, so fishing was twice a day at all hours of the day or night in all weathers. Tons of reching had to be cut every year to maintain the garth and the nets had to be kept free of seaweed and any other detritus that flows downstream, otherwise the flow of water would break them down. The posts tended to last many years, after which they were replaced.

The old cotton net used to haul the fish in was 75yd long and very heavy. When the time was right the operators walked the net around any fish that were trapped and then hauled it into shallow water. A sort of beach seine, if you must. If there were more than one or two fish it was heavy work. All salmon were sent to Muncaster Castle or sold to local hotels.

Local man Arthur Wilson once described the work it entailed and how fickle fishing can be:

> Tommy Raven and I once got sixty-four fish on a morning tide. We got two in the afternoon, then none for a fortnight. Whitebait by the hundred-weight that were served up in the Pennington Arms bar. On a couple of occasions 6ft tope sharks, mackerel that nobody wanted and other strange fish with fins like legs.

Fishing the garth finished in the late 1970s because of the decline in fish stocks and increase in licence duty. The cost of the licence was £50 in the 1960s and increased to £65 in the early 1970s. On 1 January 1976 it went up to £100 and on 1 January 1978 to £130. In 1979, an application was made by Muncaster Castle Estates for a reduction in the annual licence duty on the grounds that its value to the estate was of more significance historically than commercially. The request was turned down. Like all other garth and baulks, by the end of the twentieth century, they had fallen into disrepair. In that same way, we'll fall back into order, and proceed back to Morecambe Bay!

22

BETWEEN TWO RIVERS: 'GOING TO THE SAND'

Back in the bay, and pulling myself away from the magnetic effect that the remains of these old fishing traps seem to have on me, it was time to head north from Morecambe. Passing the signpost to Red Bank, I recall meeting the cockle pickers so many years before, when writing on the subject. Skirting the bay, we skimmed its edge at Silverdale and then came to Arnside, sitting prettily on the bank of the River Kent. Here the famous boatbuilders of the Morecambe Bay prawners, or nobbies as some call them, were based. Francis Crossfield, son of John who had a joinery shop, started the boatbuilding by launching his first boat in 1838 and the company went on to produce a host of these bay fishing craft and others, and offshoots of the family also set up business in Hoylake and Conwy over the years. These nobbies, perfectly designed for the Irish Sea waters, and found in favour by fishermen, were in use anywhere between the Solway Firth and Cardigan Bay.

Time was short so we didn't linger, moving on and eventually crossing the Kent and onto the wild and rugged Cartmel peninsula that, even today, seems a distant relation to the Lakelands to the north. Our destination was Flookburgh where, when the locals are off fishing, they simply say they are 'off to the sand'.

Flookburgh could be described as something of a throwback, existing in a past time. It has the characteristics of such when you study its buildings and layout. At the same time, it's not escaped modernity. Take away its cluster of housing on its periphery, and concentrate on Main and Market Streets, and you can almost hear the voices of a time before. People and horses, then people and tractors. The central square, with its Cockles convenience store, lies at a sort of crossroads, one road in from the north, one out to the east, one west to the sands and another south to the sands. But there's some intangible wholesomeness about the place.

I have fond memories of my first time here when writing that article on cockles, having been sent this way by the editor, Jenny, of the magazine *Maritime Life & Traditions*; the same Jenny that had allowed me to use her material on Harold Gardner. I'd walked out to Humphrey Head on a summer's evening, the last bastion of wolves in Britain, they say, and later that evening chanced upon shrimpers' carts at the bottom of the straight road south, known locally as Down Mile Road. It was one of the entry points for the Flookburgh fishermen onto the sands, giving access to the Kent estuary, the other entry point being to the west of the village, at the appropriately named Sandgate, with its access onto the Leven estuary. Later on I discovered that these indeed were not carts, they were 'chassis'. Carts are different, though even now I am not sure I know the difference!

There are two names that resonate in regard to fishing at Flookburgh in days of old: Cedric Robinson and Jack Manning, and both were born back in the fishing heydays within a month or so of each other; Cedric in 1933 and Jack slightly earlier in late 1932. Cedric was, for fifty-four years, the Queen's Guide to the Sands, a role he only packed up in 2019, the year that Jack sadly died. Cedric chose his successor, Michael Wilson, a 46-year-old fisherman from Flookburgh, only retaining his ambassadorial and advisory role. And it was Michael, and his father John, that I'd come to see.

Flookburgh is today a quiet and almost lonely village, though, as we wandered its ancient streets, it's easy to imagine how it must have been when Jack and Cedric were kids. At one time there were 100 people fishing the cockles on the sands and horses were commonplace then. When they all went out about the same time at night to go fishing, they'd go down past the houses and folk would be able to identify who was passing just from the patter of a horse's clip-clop or the squeak of a cart (as they were back then). So much so that sacks were tied around the horse's feet when you wanted to sneak out fishing and not let anyone follow you.

Some horses were kept in stables that were situated down Mile End Road, according to Cedric, in eight or so stables, so presumably that was when they were going out through Sandgate. Cedric also tells of one horrific time when there was a fire at the stables during the time when they were constructed from wood:

Shrimping by horse and cart. (Jennifer Snell)

When you bought a horse, it would usually come by rail, in a lovely horse box in those days, to Cark Railway Station. And lots of the men as bought them, they stole the halter which belonged to the rail. You know, they should have left it, but they were so well made and so strong, and John Hodgen's father's horse was tied with one of these halters, and when there was a fire and he couldn't get loose, it was burnt to death. I'll never forget it. It was just like a big cinder there after the fire. Our horse escaped. We didn't know when we went down, but it was down at the watering trough, and it still had its halter and its shank broken, but it had pulled free.

How long they'd been shrimping with horses is unknown. Some say there were as many horses as folk that lived in the village at one time! It has been said that because Flookburgh was largely inaccessible before the twentieth century, no real market grew for their shrimps, though they must have been catching them for the local market. However, it is known that hand-nets – sometimes called power or push nets, which were up to 6ft in width – were pushed through the water ahead of the fishermen. The nets were fixed to an ash hoop and a handle of pitch pine that was 8ft long, and shrimp fishing in this manner was referred to as 'putting'. The trouble with this method was that they could only go into shallow water for fear of being swept away. The shrimps were generally to be found in the fast-flowing water of the channels, and in the dykes and gutters that flow into them. As the tide receded, the channels got smaller and so the concentration of shrimps in them higher. Thus horses were employed to draw larger nets through water that was a bit deeper. The net, of half-inch mesh and fixed to the back of a high-wheeled cart, was conical in shape and referred to as a shank, as already described.

It's amazing how deep the horses can go, though the fisherman has to have a good knowledge of the whereabouts of deeper pools. (Jennifer Snell)

The horse-drawn shanking carts dragged these nets through the shallow water, bouncing the shrimps off the seabed and into the nets as they progressed along the sandy bottom. These nets, some 12in shorter, were pulled parallel to the shore. The process was surprisingly technical and reliant upon the tide, as was everything else on the sands. It was usually done downstream on the ebb, which made things easier for the horse. The weight of the catch was judged by the heaviness of the trawl rope. After stopping the horse, lifting and emptying the net into the cart, the whole outfit had to be turned around before the beam trawl was released just in case the net floated underneath the cart, thus becoming fouled on the axle and wheels. If this was attempted with the net still out, it was likely that the whole cart would be overturned by the strength of the tide.

The horses came in all sorts of shapes and types from the lighter-legged, half-bred hunter to the onerous Shire and Clydesdale sorts. The lighter animals were favoured on the outgoing trip as they reached the shrimping ground faster, thereby claiming the best fishing position. On the return journey the heavily laden cart told on the tired hunter, so that the 'tortoise' would be enjoying his warm stable while the 'hare' was still plodding home. They were all placid creatures, reliable and strong for the work, especially when they had to haul a full cart out of the wet sand. They spent many hours in the water and seldom had time to dry. Sometimes they were bought cheap at one of the local horse fairs as lame horses with an inflamed leg or foot but after working in the saltwater for a few weeks, this soon cleared up and the horse was sometimes sold at a profit a year later.

They learned fast, these horses. The newcomer was at first easily panicked in wet sand, though they soon recognised it. They had to understand the voice instructions from the fishermen while trawling, especially if an obstruction was snagged. They also learned to follow their outgoing hoof prints on the way home and could sometimes 'smell' their way back to the safe shore. Sometimes they alone were responsible for fishermen finding their way home, so that the men's lives depended on the horses. Accordingly, they were well looked after and fed on a good diet of hay, oats and bran. They were also good at swimming with the cart attached when they were out of their depth and many a horse safely carried his owner back across the channels in the sand. The bond between man and horse was very strong; it is was always a sad day when their horse finally had to be retired or, more tragically, died.

Death out on the sand was not uncommon for the fishermen. In 1857, ten fishermen drowned while returning home for Whitsuntide when their horse and cart overturned in a pool. In 1881, Margaret Sefton of Flookburgh, aged 74, was overtaken by the tide while she was out picking cockles. Again, in 1912, three fishermen from Flookburgh were drowned. There were many more tragic deaths, while we all remember the tragic events of 2004 and the loss of twenty-three Chinese cockle pickers. For, I guess, it is cockling that is what Flookburgh is best known for these days; out on the sands with the craam, a short-handled rake with three prongs used for flicking the cockles rather than raking them. They used to say the sand would spit at them when walking if there was a good bed of cockles, a bit like the patter of rain on the sand. Cockles were bagged and sent off from the station to towns all over the north-west: Liverpool, Manchester, Bradford and Bolton.

Ferguson tractors came at first, hand-starting petrol models until John Hodgen brought the first diesel one and soon the old petrol ones were a thing of the past. Today it's tractors all round, and not a horse in sight. It seems that the shape of shrimp net never changed, though they grew slightly in size between horse and tractor fishing. By the 1960s they were using a 15ft beam trawl, the mouth about 15in high. Shrimps are on the sand under the water and as the trawl comes along, it causes the water to surge up over the beam, taking the shrimps back into the net. The beam was hardwood, weighted down by iron bar, and towed behind some 30ft of warp. In more modern times the bottom beam was a 2in diameter scaffold pole. But the carts got bigger and became 'chassis'.

However, we were here for the salmon fishing and once again we've digressed and strayed away from the writing path. So, before describing my present visit, I cast my mind back and recall the time here back in about 2010 when Jack Manning was talking about his book (he kindly gave me a signed copy) entitled *It was Better than Working,* his memoirs of a life fishing, which he self-published. In it, he talks about being prosecuted five or six times for illegal fishing and this mostly involved salmon. He mentions fishing on the east bank of the River Leven in the late 1950s:

> When we arrived at Proctor's Point, to the west of Low Frith Farm, where there was always a deepish hole in the river [his father] said, 'Let's have a go at drawing just for curiosity to see if there are any salmon.' On the first pull we landed eighteen salmon and exactly double that figure with two

more sweeps. This was with an old cotton draw or seine-net, which was nearly rotten and certainly had many tears in it. I believe that if we had been in possession of a decent nylon net we would have caught hundreds of salmon on that day.

Cedric Robinson also talks about using a draw-net on the Leven probably, by the description, at the very same spot as Jack describes above, at night when he was fishing with Jack and his dad Harold, and they netted twenty salmon in one draw and had another draw, thus filling up their small rowing boat with fish. Then they'd go and hide the fish in bags under the hedge of a farmer, who knew what was going on – the poaching, that is – and Cedric also mentioned that Lord Cavendish of Holker Hall did too as they'd supply him with a fish or two, and he was happy for them to poach his river! In fact it was said that, at one time, almost all the salmon caught on the Leven was illegal!

Before 1954 it was illegal to catch any salmon on the Kent and Leven, though they used to poach salmon with a garden fork with sharpened prongs in the shallow bars on the rivers. Then they brought licences in, restricting their numbers to six on the Leven and eight for the Kent. In about 1978, Jack mentions a terrific run of salmon in the Kent when the eight licensed fishermen did really well. One day they caught 100 fish between them on a bar in one tide. For the Leven it was in the early 1970s and some fifty salmon were caught in one day on a bar on the west side of Chapel Island – thirty in the morning tide and twenty in the evening tide. And for this they used the lave-net.

I'd not realised until reading the book that they'd adopted the lave-net around the estuaries of these two rivers. Their lave-nets were similar to those on the Severn, though much smaller, and Jack writes on how he copied the design. Until he did so, they carried a fixed frame net that could be awkward to carry in a cart or tractor. He'd seen the Severn lave-netters with their folding net in the shape of a 'Y' and eventually, by watching the blacksmith at work, he managed to work out how to copy it.

During haymaking time he'd work in the fields, though, in 1966, he managed to put the pitchfork through his foot. The very next day, being off work, he saw a salmon swimming in the Leven and caught it with a lave-net. Over the next month he had one of his best years at the salmon, most in a baulk net a few hundred yards west of Chapel Island.

Tractor with chassis towing a beam in more modern times.

Of the baulk nets, he spoke of them being usually between 200 and 300yd long, usually set for flukes. The stakes, some 5ft long, of ash or hazel, were worked into the sand by some half of their length, and at about 4yd apart. They were finally hammered in with a wooden mallet to form up the sand around them.

Traditionally baulk nets were in lengths of 20 yards, laced together to allow damaged ones to be replaced. They were hung on to the stakes by knotting top and bottom cords to the end stake, then simply putting one turn of only the top cord around each stake at knee height but both top and bottom cords on every seventh or eighth and at all corners or changes of direction of the net. A spar or a thick stick of hazel is then twisted into both top and bottom cords near to each stake where the net is fastened to the stake only at the top. This allows the bottom of the net to lift except where it is fastened at both top and bottom, thus allowing fish to pass through when the tide comes in. This also stops it from rising too high and thus preventing it from falling back over the top as the tide ebbs. As the bottom of the net falls back down on the ebb it is kept on the sand by pressure from the water flowing against the spars. Traditionally baulks were set in an arc, formed by series of straight lengths with slight changes of directions at any top or bottom. These nets were often set on great flat expanses of sand so the idea was to 'surround' the fish or aim to turn them in from the ends to the centre of the net. A turnpike or small maze was formed at each end, which baffled the fish and prevented them from going round.

Michael Wilson lives bang in the middle of Flookburgh and comes from generations of fishermen. I found him just after he'd spent a morning fishing mussels off Foulness Island at Barrow. Even though he'd had a back-breaking morning, he was as affable and chatty as anyone I'd come across. You could tell he was a mussel-cum-cockle fisherman, lugging hundredweight bags all over the sands. Tall and sure in jaw, he was also built like a brick shithouse from all the hefty work, but nevertheless he had an open and welcoming attitude at the outset.

Mussels, he said, were the mainstay of his fishing at that time, though, like all fishing, things changed rapidly. Cockles, the very occasional shrimping, and catching flukes generally keep him busy. But I'd come to talk salmon:

I've fished salmon, my dad fished, my grandfather fished it. Just lave-nets. You stand in the river ankle deep on what we call a sandbar, which is a shallow part of the river, and the salmon come down backwards on the ebb of the tide, and they think 'we're going to get stuck on this sandbank' so they strike back up and because it's cloudy and sandy and they can't see you so you run down in front of them and basically just scoop them up. You just stand on the sandbar and watch and you see there is a mark in the water. There'd be several spots in the river as the bay went out, maybe further up at Arnside, or across at Grange. We can go right up almost to the viaduct, to the pier at Arnside to be precise, and then right down right into the estuary. The same in the Leven, again just below the viaduct down towards Barrow, wherever it forms a sand barrier. The River Winster used to be fished. You could always tell a Winster salmon to a Kent salmon, it looked different. If you caught one and it looked a bit stained it was a Winster salmon. There used to be some salmon in the Eea that comes out at Sandgate and we used to get a few fish there. But there's none now in either the Winster or Eea.

Michael talked about the stream-nets, bag-nets, drop-nets and boat-nets. They still set stream-nets occasionally to catch flukes and these are nets on posts that fish with the tide, with boat-nets simply being bigger versions. Drop-nets are just stake-nets, using monofilament nets that catch fish in both directions.

I was keen to see his lave-net to compare it to those from the Severn. There were two in his garage, one being his and the other his father's. They were much smaller than the Severn versions and much more basic in that one solid piece of oak made the handle and two broom handles made the wings, so that, when extended, two iron cradles swung out, one each side, to hold the two wings in place. The net was strung between so that the distance between the ends of the wings was about 30in. The annual licence tag from 2008 was attached to his dad's and 2012 from Michael's, showing when they last fished. The licence in 2012 was £150 and, due to the lack of salmon, fishing was just not worth continuing.

The short folded net was kept in a fish box aside the tractor and he showed how the end slotted through a purpose-made hole in the fish box.

> You'd be out shrimping and you'd see a salmon and grab the lave-net quickly. Before, the handle used to get in the way and by folding it, it's easy to tuck it away in that box. That was why Jack wanted to fold his up. It must have been those two iron cradles that he 'invented'.

We chatted briefly about what it meant to be the Guide to the Sands and how he can sometimes be leading 500 paying people across the Kent. Another fellow, Raymond, guides across the Kent from Sandgate. 'Lovely fellow, but he's so laid back, he's like a plank,' says Michael, though he could have been describing himself.

Next stop was the family land a mile along the road on the edge of the village. This they called Orchard Field, a couple of acres, which used to be a fully growing market garden. Today only part of the ground was growing rhubarb, potatoes and beetroot as the market trade had disappeared, and I'd seen an honesty table with those crops on outside Michael's house. They'd sell produce at the Barrow market, as well as from the shed, until they stopped going to the market in 2019. The supermarkets, you see, that took that trade away until it just wasn't worth the effort of the twice-weekly journey. Something his dad John had done since the 1950s and his father before him. All sorts of veg – cucumbers, carrots, leeks as well as the rhubarb and spuds still growing. Fish too, and pots of herbs and veg like lettuce, even the odd flower such as pansies or whatever. It was once a thriving business, hopping on the train with the cart, selling. It was the norm and up to maybe thirty Flookburgh families once did it. Now it's all gone, ways of life destroyed by the corporate.

Sadly the land was for sale and Michael reckoned they'd end up having twenty-four houses built. Inside the shed was the oil-fired shrimp boiler and the last vestiges of this being the family's market garden-cum-fish shop. Various cockle and shrimp riddles, as well as other remnants of fishing, sat around alongside cards with the prices of various veg written on. Next door they'd nailed all the salmon licence discs on a beam over two tractors, dating back half a century. There seemed to be tractors everywhere and, when asked, Michael admitted they had fourteen old ones.

His dad's shrimping chassis lay alongside a bed of rhubarb, with two 14ft shank nets atop. I asked why only 14ft when they are allowed 30ft of net, when, conveniently, his dad John turned up. Michael relayed the question to him, who replied that it was lighter than a 15-footer. He said he'd have preferred two 12ft trawls, given the choice.

The whole place eluded an atmosphere of sadness, as if it was surrendering to the inevitable. Here we were seeing the complete end of an era. Not just the loss of generations of traditions of working both the fish and the fields, and selling both produce to eager locals instead of our food attracting food miles by being carried further and further around the world, but something much more. This place, with its sheds, tractors and vegetables is something tangible, unlike business driven by internet sales and fast transport. Looking around it's easy to see that these fields once thrived and kept the various generations of families satisfied. Now it will simply produce money from house building and turn yet another echo of the local salmon, cockle and shrimp fishery, and much else in today's fishing industry, into the melting pot that is history. Finally, as we go to print, I heard that the land was sold. Soon new housing will arise, as it's doing all over Britain, but the housing will never last the test of time as the fishing has.

23

SOUTHERN SOLWAY

I'm not an avid Lycra-shod cycler but when it's flat I can see the various advantages of this mode of travel. Thus, one autumn for a few days, daughter Ana and I spent time on our bicycles traipsing along part of the southern edge of this coast. Now the Solway Firth can be said to stretch from St Bees Head, just south of Whitehaven in Cumbria, to the Mull of Galloway, on the western end of Dumfries and Galloway, but don't get excited. We did not cycle that far south. If the Firth is the border between England and Scotland, between Cumbria and Dumfries and Galloway, then I was happy to keep well within sight of the Dumfries part of the coast where it consists of unexciting beaches and salt marshes, havens for bird watchers.

From the point of view of our two fish, it's probable that Whitehaven, Workington and, especially, Maryport were ports where quantities of herrings were landed. Further north, Silloth, with its man-made harbour, built in the 1850s to replace Port Carlisle as the deep-water port for Carlisle, was a haven for fishing vessels. In fact, it is only Silloth and Annan that offer any shelter within the upper reaches of the Firth. Today, although Silloth became a northern Victorian seaside resort after the arrival of the railway, the harbour is often empty, and local shrimps are the only delicacy, sold as Solway Shrimps, either fresh or potted. The other attraction is the East Cote lighthouse, built in 1864. It's a wooden structure built to be mobile on a short rail track. It was the rear of two leading lights indicating the approach channel to the port. In 1914, it was fixed in this optimum position and at that time had a small cabin on a plinth below the tower for the keeper. The light was automated in 1930 and rebuilt in 1997.

Nevertheless, our shoreside perusing began here, and just along the coast is Skinburness, a tiny hamlet dating from 1100, a time when nearby monks cleared the marsh and scrubland for farming. Soon, as the information

The Highland Laddie Inn was unexpectedly closed.

sign told us, a port was established, exporting wool, fish, grain and salt to Ireland and the Isle of Man.

Salt, of course, was once vital in the preservation of food: fish in this case. Around here are local place names such as Saltcoats, Saltpans and Salta, indications of a serious salt-making industry from the natural salt beds. Remember the Romans were here and how they preserved their seafood.

We continued around Moricambe Bay, often referred to locally as Hudson Bay because of the wrecks of aircraft from the Second World War Silloth aerodrome. Silloth lore accredits local fishmonger, Joe Lomas, as having rescued many aircrew from ditched Lockheed Hudson aircraft, and many are thought to have crashed hereabout. One figure puts the number at sixty-four. Meanwhile, into the bay flow the rivers Waver and Wampool on their journey into the Solway Firth. We continue around, past Anhorn with its radio aerials transmitting secret orders to submarines as well as time-keeping.

We came to Bowness-on-Solway, where the Romans had chosen to place the western end of Hadrian's Wall. The village is also the southern end of the rail bridge across to Annan, on the Scottish side. Built in 1869, it was an iron girder viaduct just over a mile in length and remained in place until being demolished in 1934. Some say the reason was to stop the Scots, whose pubs were closed on Sundays, walking across to the more liberal English side, and returning in a less than sober state and occasionally falling into the Solway and drowning.

We cycled through the village, all quiet on this early Sunday morning, and continued on to Port Carlisle (originally a small fishing village and port dating from 1819 called Fisher's Cross), where the entrance to the old Carlisle Canal once stood – if something like a canal basin can 'stand'. This short 11-mile canal, perhaps one of the lesser known in Britain, opened in

1823 to allow vessels to sail into Carlisle instead of fighting the river up to Sandsfield and carting the goods into the city. However, it was a latecomer to the world of canals and by the 1840s, railway mania had gained ascendancy, so that in 1852 the Canal Committee resolved 'that it appears highly desirable and expedient that the Carlisle Canal should be converted into a railway with as little delay as possible'. This proceeded, wiping out much of the canal, although parts are still visible. With the opening of the viaduct across the Solway, this route soon faded into obscurity. Thus ended the short, busy chapter in Port Carlisle's history.

Following parts of the canal that were visible through the trees, we pedalled east, and again alongside the shore with the signs denoting flooding of '2 feet if the water reaches here'. Thence to the village of Glasson. Here was the Highland Laddie pub, once a hostelry with haaf-net tours and where we hoped for some refreshment. I'd found this place on the internet advertising these tours and I'd emailed but had received no reply. So I had decided to come in person. But, oh dear, the place was boarded up. We turned around and retraced back to Bowness, where I'd seen a pub. There I discovered that the Highland Laddie's publican, Mark Messenger, had moved to a pub in Carlisle.

Over a pint at the bar, always a good haunt to learn, I heard much more about haaf-netting in these parts. The barman introduced me to a fellow whose name I forgot, and who hadn't actually netted but was a mine of information.

Classic photo of haafing in the Eden estuary, according to the fellow in the pub.

'Yes, Mark Messenger and his son Ben have licences to fish,' he told me. 'But, although they ran the Highland Laddie for a number of years, they left it only a few months ago.' No wonder they hadn't replied to my e-mails, I thought.

I asked him to explain about the fishing, and the fellow continued:

There were once getting on for 250 licences here on this side of the Solway. Now it's fifty, though they can't land the fish this year. Bit of a waste of time in my book but there are those that still go out. Tom Dias, he was the man you should have spoken to. He had netted since the 1980s. Always said he had learned from the best. Ernest Percival was one name I remember. They said his word was the law. It was after the war when food was rationed that fishing gave the men extra food at no cost so they were out there in force. But they were always democratic in how they fished. In their order, like. They called it drawing the 'boak', the 'boak' being the line of haaf-netters fishing. One of them is the 'baggie' who does the draw, so-called because they used to draw from a bag. He turns around and stands with his back to the men. The others bring out their priests and push them upright into the sand, then each man chooses a different stick as his lot otherwise the baggie might recognise one. The baggie comes back, and pushes over the sticks, giving each a number. Whoever gets number one can choose his position in the boak. Number two then makes his choice, and so on. Then they go and fish.

I bought him a pint!

They used to go down four hours before high water for the draw. But some got greedy and went where they wanted to. Others simply turned up late and went to join the end of the boak. Then the arguments started. But for most of them, the camaraderie was always there. It was only a few that didn't keep to the unspoken rules. But then the arguments spread into those anglers who reckoned the haaf-netters were taking all the fish as they weren't getting them further up the Eden. Then folk came to watch them and reported them for under-reporting their catches. Big long-range cameras behind bushes, court cases. See haafing used to be the job of the working-class fellows keen to add a bit to their earnings. Then the posh folk got upset. The Eden was the salmon river they always said, while the Esk was best for sea trout. Fellows haaf-netted up the Eden and even some seine-nets in my time. So the fellows went to court accused of fiddling their tax by saying they'd only caught so-and-so salmon but the posh folk, with the camera evidence, proved they were lying. They got fined because they were working folk, you know, in their hats and anoraks in court. The judge was probably a rod fisherman and made them pay. Didn't appreciate the fact that they didn't respect the court. Now, if you don't tag them before you leave the beach you can get into trouble.

Then he led me outside and pointed to the haaf-net leaning against the wall of the pub. How I'd not noticed it before I can't understand. Maybe the exhaustion of pedalling up the hill! It seemed slightly shorter than the Lune nets, and certainly not as high, though the base of one of the outer sticks did look somewhat rotten. The net was not there and when I pointed this out, my fellow replied:

There's been trouble that way in the past too. Nets being cut. I remember Tom Dias telling me his net was stolen once. Four were cut with a Stanley knife all at one time once. There's always folks' opponents, always will be. But then there's only the few these days that do it. You can see them in summer, out on the sands. Same as you can see the fellows across the river on the Scottish side. They fish with their haafs in the same way. And they had stake-nets on the sands, which weren't allowed this side. You need to go and see them at Annan.

And with that we said our farewells and clambered back onto our bikes for the ride back to where we'd left the van. As we cycled the last mile along the road lining the salt marsh shore, I looked across. Criffel, the mountain, always seemingly watching over the Firth and always omnipresent, had suddenly gained a cloud and was almost lost to view. Was this, then, a foreboding that haafing on this side of the Solway was about to disappear for good?

PART 4

SCOTLAND

24

A HUDDLE OF HAAF-NETS

Of course we went to Annan! It was a gloomy, but dry, sort of early evening when we, the kids and I, arrived at what we thought was the fish house belonging to the Annan haaf-netters, as described on the phone to me by Barry Turner, who is the secretary of the Annan Royal Burgh Fishermen's Association. We'd crossed the border into Scotland with a cheer arising from the kids in the back. Now the ownership of the fishing rights was different. In Scotland they are heritable rights held by individuals or groups, or the Church in some cases and, here, the town. They used to be owned by the Crown but, over the years, they've mostly been conveyed to individuals by written Crown grants for whatever reason. Today they can be bought, sold and leased, with the fishermen either being owners or tenants.

We found the spot that I thought Barry had described down a short track and just above the foreshore on the east side of the old railway viaduct across to Bowness-on-Solway, and which itself is to the east of the estuary of the River Annan. It was a small building with a corrugated-covered pitched roof and a red door, and looked just the right size for a fish house. The fact that several haaf-nets – a Huddle of Haafs, as they say hereabouts – were around the back, some among the trees and two leaning against the rear wall, seemed to support the supposition. The tide was well out and the sandbanks of the Solway Firth in abundance.

We'd been told that we could park up for the night here and so I lit the firewok and grilled up some food while a combine harvester slowly made its way around the adjacent wheat-coloured field. We were meeting Dave Nash who, in the end, didn't arrive till 2 a.m., by which time we'd fed and were fast asleep. 'Traffic on the M25' he had texted!

Waking up early, I'd been listening to the rain pounding on the van's roof on and off during the night so I wasn't surprised to look out onto a pretty miserable day. After all the good weather we had experienced to date, it was somewhat unfortunate that it seemed to have turned for the worse. Dave was already up and about with his dogs as he'd slept in his car rather than pitching his tent late. It took somewhat more persuasion to get the two kids out of bed as we'd arranged an early meet with the fishermen. However, it turned out that this wasn't their fish house, which was actually a larger building back, just across from where we turned off the road. We'd simply parked up by a farmer's shed, but what a glorious spot it was! Thanks! Nevertheless, we met up as agreed at 7.30 a.m. – Barry, his brother Tony and John Warwick – who'd kindly brought waders and oilskins for me, which I donned. Luckily the rain had stopped but heavy cloud persisted.

Before long we were off down the beach, again with the tide well out. The haaf-nets are carried on the shoulders of the men, as we'd seen before at Sunderland Point. Here the frames of the nets appear to be called haaf-backs, though are the same as those from the River Lune and at Bowness. Fishing normally starts before the ebb turns to the flood, which can mean five hours out in the water, though, on this occasion, the tide, although way out, was already on the way back up. Into the water they went, and I followed to photograph. The kids, I could see, were still gingerly stepping down the beach, through the mud, in sandals.

They stand in a line – the 'back-o-men', as they term it, the order of the line being determined by the ancient 'Casting the Mells' where they throw down their mells (the sticks used to kill a fish) and one is eventually chosen by 'kicking out' and he takes the number one position. A bit like the fellow in the pub at Bowness had described. Others are chosen by the order of mells left in the circle. However, with only three fishermen, this didn't happen, though, with thirty current licences issued to the haafers, it is sometimes necessary. Barry told me later that only twenty-seven had been taken up this year due to a lack of young ones being interested. There were once forty-seven licences but the haaf-netters themselves had voluntarily agreed with the council to limit them to thirty.

'Forty or fifty years ago licences were highly sought and you had to have a good relationship with a councillor to "speak" for you when licence applications were being considered,' so Barry told me. 'But they can't stop us fishing. The town was given the Royal Charter in 1538, granted by King James V, and is believed to be the re-erection of one granted much earlier.'

Off we go with the Annan haafers.

It seems the charter was granted to Annan in recognition of its loyalty to the Scottish Crown because I read that the town:

has often, at divers time, been burnt and destroyed, and the burgesses and inhabitants of the same, in times of peace as well as war, been plundered and slain by our ancient enemies of England, in protection of our kingdom, and have often placed their lives in extreme jeopardy, in resistance of our ancient enemies aforesaid, and in defence of the limits and boundaries or our kingdom, opposite those parts of England ...

Bloody English! Mind you, I can't imagine the Scottish Government being so loyal in this way these days even with such English behaviour! Nevertheless, these rights were again re-erected with another Royal Charter in 1612, granted by James VI after the 1538 Charter was destroyed when Annan was once again sacked and burnt by raiding English forces, this time led by Lord Wharton in 1547, the same year he got married (the two things don't seem to go hand in hand!). He was obviously a bit of a lad, this Wharton chap:

knighted in 1545, the companion of Mary I of England, named 'Master of the Henchmen' and he became the local MP at various times before eventually being imprisoned in the Tower under Elizabeth for a while. A chequered and doubtfully useful history, I'd say.

Anyway, there I was, not wishing to be associated with such an English invader, standing in water up to almost my chest, photographing these folk. They stood facing the onslaught of the tide, a wee bit of wind whipping up wavelets, eyes searching for any signs of salmon.

After a while, Tony suggested I take his haaf, which I did, and he showed me how to take a hank of the mesh of the net, threading fingers through at least two meshes side by side a couple of meshes lower than the beam to feel for tugs that denote a fish. As a 'new start' as they call beginners, I probably had 'wooden fingers', not being adept to feeling these tugs. They've a whole glossary of terminology such as 'going for a swim' for someone who has either walked into deep water or, more seriously, got swept away.

Sometimes they floated away, holding onto the haaf for safety, and sometimes fetching up on the English side, where there were once plenty of haafers working, as we've learnt. One of these would then proceed to drive them home. The same worked vice versa. I guess they had to tie the haaf-net to the roof of the car! Today, with a much lower number of haaf-nets on the English side – some forty according to Barry – the chance of ending up and being seen by one would be much slimmer and I didn't fancy the challenge!

Still, the force of the tide against me was much stronger than I'd have believed. Being unable to hold the beam against the current was 'cannae haud it', and that was fully understandable. I also had to be careful not to stand on the net, which was flowing out either side of me like two poke or pocket nets, otherwise I could topple over, get entwined in the net and suffer serious consequences. It would definitely 'be over the top' for me, with water down the waders. And I certainly didn't fancy a swim over to the opposite side, which looked a bloody long way away in the haze.

The real knack is in spotting the fish and knowing what to do if you get one in the net. If you do feel one, then the net is quickly lifted as otherwise the salmon turns around and swims out. Some have their own way that means stepping back with the current and flipping the sudden slack of the net under and over the beam, to close the neck of the net. As I considered this, as well as the comfort of standing in the water for up to five hours, I decided I'd

But we only caught three flounders.
My son Otis showing off the catch.

not make a good haafer! So, after some time, having felt nothing, I handed the net back to Tony and reclaimed my camera from deep within my clothing where I'd hoped to keep it dry even if I did take a stumble.

As the tide rose, we had to head inshore, and the pace of that rise was pretty fast. Either that or it was the time flying by. One minute the water was up to the knees and then, before you know it, it's groin high – 'baw deep', and then, if it's dripping into the waders 'over-the-top'!

I guess we must have been out a couple of hours before they'd decided they had had enough. With a catch and release in force, the only fish they'd netted was a grilse that was instantly released. Oh, and three flounders. Next thing we were plodding back up the mud, to the shed I'd thought was their fish house. The nets were 'huddled' and we chatted a bit about the legalisation.

Fishing was allowed here from the beginning of May until 9 September, not at weekends, and, unlike their neighbours on the English side, they can fish at night. As John Warwick said to me:

> Legislation on the English side of the Solway does not allow haaf-netters there to fish during night time but our legislation does. It is just that now all salmon have to be put back, there is no incentive to go night fishing (unsociable hours) and the fishermen pick and choose their tides during daylight hours.

Their area to fish is designated as stretching from Waterfoot, where the River Annan emerges into the Solway, to the Altar Stone, a glacial stone that is way east on the edge of the sandbank, and from the foreshore to the middle of the main channel. On the other side, the English haaf-netters fish up to the middle, though there were various reports in the nineteenth century of how these men infringed upon the Scots side.

In Scotland, all the rivers and estuaries are categorised by the state of their salmon and sea trout stocks, one to three, the latter resulting in a mandatory catch and release system. If you look at the river grading list given out by Marine Scotland, you'll see that the majority of Scottish rivers are category 3, although four of the famous salmon rivers – North Esk, Spey, Tay and Tweed – are category 1. The Annan and area (which includes the haafers) has a '3' rating and is therefore under the mandatory catch and release. However, in 2016, Scotland's First Minister declared that the Annan Haaf-net fishery would be designated as a Historical Fishery and thus the fishermen are presumed to be allowed to continue to practise this age-old method. Why those in power in Wales haven't seen the light and designated their traditional salmon fisheries in the same way is beyond me.

But the surprising fact is that if you consider the numbers of salmon caught in Scotland, a pattern emerges. Take the 2008 figures for instance. The 'Fixed Engines' landed 21,016 salmon, the 'Net and Coble Fishery' 8,107 salmon, the 'Rod & Line, caught and retained' 37,929 fish, and, finally, the 'Rod & Line, caught and released' 46,069 fish. Thus, the anglers caught some three times more than the commercial fishermen and retained almost 10,000 salmon more. By 2009, the figures were 8,206, 4,648, 23,690 and 48,136 respectively, which shows a dramatic reduction in commercial salmon landings and only a slight rod and line reduction. Nearly a decade later, in 2018, after all the conservation measures aimed mostly at the commercial fishermen, unsurprisingly some 91 per cent of the total annual reported catch was accounted for by the rod fisheries. The net and coble fisheries comprised 9 per cent, with fixed engine fisheries accounting for around 0.3 per cent, the latter being accounted for by fifty salmon from the Annan haaf-nets and sixty-nine by other Solway haafers who are generally working the area around the lower reaches of the River Nith. However, as it was pointed out to me, the haaf-nets are not fixed engines and if Marine Scotland has included haaf-net catches under the heading of 'fixed engines', then this is either administrative convenience or ignorance.

Thus it is easy to see how the anglers are getting the lion's share of fish. But haaf-netting hardly hurts anyone. John Bedwell Slater, clerk to the Eden Fishery Board, although referring to English half-netters, observed in the 1896 *Report of the Commissioners on the Fisheries of the Solway Firth* that:

haaf-netting fishing I do not think anybody has anything to say against it. It is a perfectly harmless thing and everybody is welcome to anything they catch. It is hard cold work. They do not sweep them [salmon] out in hundreds.

This was reflected a century later when, in 1989, a time the haafers were once again feeling somewhat threatened by extinction, an article in the *Salmon Net*, the journal of the Salmon Net Fishing Association of Scotland, commented:

With all the problems of salmon management which Scotland faces, to spend so much time and effort in the elimination of an ancient, and legal, method of fishing, is little short of Neronic!

This holds true as much today as it did back then, and could be alluded to include the whole of the UK fisheries. Unlike the Annan fishery, many other traditional fisheries haven't been recognised for what they are: part of our historical past or, in other words, our vernacular culture. One can only hope the Scottish Government keeps its promises.

And just to be sure about that last comment, I decided it prudent to delve into the past. Although it is assumed that the haaf-net is an influence entrenched by the Viking incursions and their subsequent settling in these parts, there's nothing concrete. That it was a method both on the fringes of both Morecambe Bay and the Solway Firth, similar in tidal and topographical detail to some extent, is not coincidental.

Although there were various mentions of fishing in the sixteenth century, it wasn't until 1612 that the first mention of the haaf was made in the charter that restated the rights of the Royal Burgh of Annan:

by the mire (or well) of Grekane to Merebeck running to the Sea and from this to the Altar or Otterstain within the Firth of Solway and from thence to the foot of Annan Water, shore and sand and so from the foot of Annan Water to the Northburnfoot, with the Fishing of salmon and other fish by boats, netts, haffs, cowps and from the Altar or Otterstain to the foot of Annan Water.

The fisheries of the river and nearby area continued to be mentioned in various papers from the Town Hall records. In the sixteenth century, various orders were made and the fisheries organised to benefit the townsfolk by way of the Common Food Fund. In about 1733, town ratepayers were said to be entitled to fish with haaf or poke-nets, and if they were not fit to do so personally, then they could employ or appoint someone to fish for them. In 1740, it was reported that haaf-netting was to take preference over any poke-nets:

> To the Regulations formerly made in that behalf in the fourth place that no person or persons are allowed to set any poke-nets to the prejudice of the Haffing either in flowing or ebbing before Lambass yearly without liberty from the Magistrates intimated to the inhabitants by Luck of Drum in the fifth place that the act of Councill made on the 23rd July 1740 regulating the fishing with poke-nets do continue and remain in full force. In the sixth place that the transgressors of the regulations appointed by this present act shall incur and forfeit a penalty of ten pounds Scots money for each transgression toties quoties. To be recovered by prosecution before the Burrow Court at the instance of the Procurator Fiscal and the fines and penalties to be applied to such uses as the Magistrate shall direct and lastly the Magistrates and Council do continue John Anderson and John Johnston water baillies with the same powers and privileges granted to them by the aforesaid act of Council of the 19th of June 1733 and is hereby declared that this present Act is to continue in full force until it be altered or repealed by a subsequent Act of the Town Councill (signed) Bryce Blair Procurator Fiscal and H Currie Baillie and George Johnston Baillie.

So haafing was obviously given some priority and recognised back then for what it is: a passive way of fishing that has minimal effect on salmon stocks. Almost sixty years later, in 1797, it was written in the Old Statistical Account for Scotland that, 'a curious species of net is used here for taking salmon, both at the flowing and during the ebbing of the tides'. Salmon were said to be plentiful. The description can only really relate to the haaf because of the mention of both tides.

Today, they still celebrate the Royal Charter in July, when the boundaries of the Royal Burgh are confirmed when a mounted cavalcade undertakes the Riding of the Marches. Entertainment includes a procession, sports, field displays and massed pipe bands. These events occur in many of the Lowland border towns and in Langholm, I learned, they go as far as offering a salted herring. It seems that for the Common Riding, a salted herring is pinned to a bannock (a variety of flat quick bread), which is fastened to a wooden platter by a 'twal-penny nail' and flourished aloft on a pole. The bannock symbolises privileges of the Baron under the obligation of Thirlage (the feudal servitude under Scots law restricting manorial tenants in the milling of their grain for personal or other uses) and the herring may be symbolic of the right of the baron to the fisheries of the Esk. The things you learn!

And so, as I fought to remove the waders (which were slightly small for my size 12 feet!), watching the tide flood up the foreshore, I felt sad for once again I'd experienced the beginning of the end of yet another tradition. Dave and I, and the kids, said our farewells to Barry, John and Tony, and they drove off. The kids and I sat a while on the metal-sculptured bench against their fish house wall facing the sea. The back was a beam of a haaf and there was the odd salmon swimming on the ends of the bench below. Alongside that was a similar metal sculpture of a haaf with an information board attached to the 'net', and alongside that an old wooden haaf which had one end stick longer than the other for fishing the side of a bank or breest, as it's called. While the old haaf will soon fall apart and fade back into the ground, the metal sculptures will presumably last the course, a place for the subsequent children of the haafers to sit and think about the living history long gone.

In the book *Annan Haaf-nets*, which aims to promote the cultural and historical importance of haaf-net fishing (full of personal reminiscing and strongly recommended to anyone wanting to learn more on the subject), and was written by John and Tony; on the penultimate page they write that the haaf results in 'practically a zero mortality rate' when fishing catch and release, whereas angling's mortality rate is, according to government statistics, 10 per cent. Whereas the anglers are allowed to continue to kill thousands of salmon by ripping hooks out of the mouths, the haafers account for zero deaths. Once again we see that it's not just a matter of conservation, as the eternal excuse is. If only the haafers were allowed to retain one fish per man per season, this alone would help attract newcomers into it so that the knowledge and skills can continue to be handed down through the generations. Without this haaf-netting will, no doubt, die out completely.

Barry has been fishing the Solway since 1976 after moving up from his native Manchester. That was a time when, he says, fishing at Gretna was like stepping into the Wild West. Many of the fishermen were virtually full-time haafers and used every dirty trick in the book to gain an advantage over other fishermen, including sabotaging other fishermen's equipment. As soon as darkness fell, what rules there were ceased to exist, as did recognition of the fishing border, which separates England and Scotland! In those days, fish were abundant and anglers were catching as many fish as haaf-netters, so bailiffs seldom appeared. As a result, a kind of pleasant anarchy existed. He says his favourite form of haaf-netting is fishing flood tides. Half an hour of the ebb tide and then fishing the flood tide, ending the session by holding his beam by the end would be his ideal day out. He loves fishing shoals – where the water pours in fast across big areas of shallow water – but only when he can get a shoal to himself. One year he and his brother Tony discovered that it was possible to shoal fish in Seafield at four-hour ebb. They had it all to themselves for weeks, even hiding fish in the sand to deter prying eyes. That was fun, he told me, though that situation seldom happens as there are lots of guys with better eyesight and faster legs than he has. So, as secretary of the Annan Royal Burgh Fishermen's Association he should have the last word, which echoes what I've already said:

Regrettably, haaf-netting is on the verge of dying out. We are now not allowed to retain any salmon we catch. The Scottish Government has forcibly shut down the town's stake and poke-nets. By doing so the government has deprived Annan Common Good Fund of an annual five-figure sum it received from net licence revenue. This is patently unfair. The government must offer compensation. If they continue to refuse to do so, Annan residents will miss out on this sum forever. The remaining thirty haaf-netters must be granted a small quota of salmon, otherwise a unique activity and part of the town's identity which dates back to at least Viking times, will be lost forever. If the government refuse to permit us a quota I believe it will be guilty of both cultural and historical vandalism. Any Scottish government should defend and preserve Scotland's heritage, not destroy it.

25

STAKING OUT
THE SOLWAY

We have already noted how Morecambe fishermen introduced whammel-netting into the Solway in 1855 when the Woodmans and Willacys arrived. At first this seems to have been an unlicensed form of fishing and mainly concentrated on the English side of the Firth as, according to the Annan Act of 1841, use of the whammel-net was regarded as being outlawed. In 1872, the River Eden Board started issuing licences for whammellers and within three years, any boat working a whammel-net was forced to obtain a licence. In 1880, it was demonstrated that the Annan Act only referred to fishing down as far as the low-water mark and therefore fishing between here and the middle line of the channel was legal. Nine years later there were thirty-nine licences issued and, although the Scots fishermen resisted the whammel-net to begin with, by 1895 all the licensed fishermen were Scottish.

Between about 1935 and 1938, the Society for Nautical Research commissioned Philip Jesse Oke to measure up various traditional coastal sailing craft from around the British shores. One of these was the Annan whammel boat *Dora*, built in 1900 on the waterfront at Annan. She was 19ft 3in in length, had watertight compartments under the side decks and was iron ballasted. She was rigged with one lugsail and a jib, although the mast was usually lowered when whammelling. This type has been said to be a river and estuarine craft but, in comparison with the smaller open whammel boats belonging to Tom Smith – *Sirius* and *Mary* – they are obviously used to more exposed waters. They might have been confined to the upper reaches of the Solway Firth with its shoals and fast currents, but, with their extra couple of feet in length, side decks and watertight compartments, they appear more used to foul water than Tom's boats. The belief was, I learned, that whammelling was more effective in rough weather.

Whammelling is, as previously mentioned, a form of drift-netting, the net being smaller than the traditional herring drift-net. The 'net and coble' fishing for salmon, on the other hand, is different in that it is vital that the fishermen do not let the net drift. In this case the net is loaded onto the coble while one end remains on the shore with a fisherman. Once the operation begins, he must keep the rope in motion through his own exertions. The coble moves out into the river, the fisherman shooting the net as he goes, rowing in a semicircle back to the shore so that the two ends of the rope join. Hauling then begins and here it is important to keep the net moving. It must not become stationary or begin to drift as the salmon are persuaded into the landing ground, where they end up in the bag, encouraged by splashing the ropes as necessary. Thus the salmon is not enmeshed by the gills, unlike the whammel-net. In actuality, it's basically the same method as we were following out on the River Severn with a long-net.

Annan definitely seems to have been a Mecca of fishing in the nineteenth century after the Lancashire fishermen settled there in the 1850s. In 1896 Annan had, among a host of other fishing craft, twenty whammel boats. Although shrimps seem to have been the main catch, the fishing declined as more modern boats moved down to Kirkcudbright. However, in 1962, whammelling was outlawed in Scotland, though the practice continued across in England until the last whammel-net was bought out by the North West River Authority about twenty years ago. Strangely one whammel boat was based in Annan in 2000, though it is not known whether she fished over the centre line of the Solway.

Although whammel-netting in England only ceased recently, back in 1962 on the Scottish side, the use of a sparling-net, a small 100yd fine mesh net fished as a draft-net as a winter alternative for whammellers, continued for some time more. Sparling (otherwise known as smelt) are a small pelagic fish (6 to 9in long) that taste and smell like rushes, according to Angus Martin in *Fishing and Whaling*, and it was once widely caught in Scottish rivers and inshore waters but now only in the rivers Cree, Forth and Tay.

Unlike the whammel-net, the stake-net was one of two fixed engines used in the Solway Firth, the other being the poke-net. These, like haaf-nets and poke-nets, were once in use from Newbie all the way up to Gretna, and it was said that salmon went by train to Billingsgate as a daily occurrence. According to the 1824 observations of the salmon fishery by an anonymous writer, the stake-net is said to have been developed by two brothers, William and James Irvine:

On the extensive flats or sand-banks in the Solway Frith, large excavations are made by the eddies of the current, which, at ebb tide, form on the banks large pools – or lakes, as they are termed by the fishers. At these lakes, the fishers erected what was at first called a tide or floating-net, in consequence of the net being so constructed, that it was the operation of the tide itself which secured the fish. It consisted of strong and coarse net-work, the meshes of which were ten or twelve inches in circuit, placed along the margin of the lake and surrounding it on all sides. This net-work was fastened to stakes driven into the banks, at considerable distances from each other; and at various places in the lower or flood side, it was so constructed as to open and shut with the current. These places, again, were kept open by the flood tide, so that the fish, during the flood, were allowed to go freely into the net; but when the current of the tide changed and took the opposite direction, the loose net-work, pressed by the receding water, was closed, thus forming a complete enclosure, in which the fish were detained. And as the tide ebbed, they sunk down into the lake, where they were caught by the fishers, at low water. Such was the origin of what is now called the stake-net.

Unfortunately, there's no date within the passage, but we know they were in use much earlier than 1824 since, according to a booklet written by John Warwick with archives from Dumfries and Galloway Council, a proposal was made that suggested the formalising and letting of stake-nets in 1758. However, there is some confusion, as it has also been said that the first stake-net set was at Newbie in 1788. I wonder whether there has simply been some misreading or typing error when distinguishing between 1758 and 1788. An easy mistake and either date could be accurate. I'm not sure it really matters!

The stake-nets were then let to individuals for a varying number of years, although a fishery manager was employed to run the stake-net fishings at other times. They were always to be run for the benefit of the town. By the 1820s, they were leased to a jackman:

That the foresaid Jackman shall have no right or under any pretence whatever be entitled to place his salmon engines as to interfere with or injure the haffing fishing on the Stennarhall briest or poke-nett fishings on the inner or southern banks of the said fishing grounds which are hereby expressly reserved to the community of the said Burgh as usual; but shall occupy and possess.

Inside the Snaab star-net. (John Warwick)

It's an interesting use of the word 'jackman' as usually this is a 'retainer of a nobleman', which suggests he was in the pay of someone higher up the social scale than the town provost. Nevertheless, there was interference between the different fisherfolk working different methods. An interesting complaint from other fishermen against the stake-nets of Jackman Robert Carlyle was recorded in 1858:

> We as under stent holders of Annan petition your honours for protection to the rights we hold for fishing on the Sand Rigg. We complain that Robert Carlyle now taxman for the stake-nett fishing Seafield has of late years infringed on own boundaries on the Sand rigg and at this time has put a fly nett in nearly opposite the Whinnyrigg house which reaches from the shore nearly to the top or crown of the Sand rigg which he has no right to any portion of claim whatever infact the Saud Rigg has been cut up in consequence which debars us of our liberties entirely. We begg therefore that your Honours will take a view of the matter and trust that you will prevent the continuance of such unlawful steps taken by Robert Carlyle and grant to your petitioners their just clauses to the Sand rigg which your petitioners ever crave.

Inside the Snaab star-net. (John Warwick)

However, it is unclear as to the eventual outcome, although the council did go and inspect the stake-nets and decided that Carlyle was in his right to fish there and 'to intimate that the stent holders have complained to the Council that he is trespassing'. The stake-nets have always been the main source of commercial income for the Annan Common Good Fund but since 1870, the haaf and poke-nets have also contributed.

A good description (if not long-winded) of stake-nets comes from *A View of the Salmon Fishery of Scotland* by Murdo Mackenzie (Edinburgh, 1860):

> The fishing-apparatus is formed of a long range of stakes carried out from the shore sometimes a mile into the water, that is, to the low-water mark, with nets and traps affixed to them: the whole forming a barrier of the most formidable description in the course of fish, and which, it must be evident to all, must not only break and scatter the shoals as they come on, and thus most materially hurt the fishery, as regards the public, but which may be so multiplied on the coasts, as to ruin the river fisheries entirely ... When a shoal meets with a stake-net, some of the fish are caught in traps, or cruives, or what is called its chambers, others start off; in short, the shoal is broken and dispersed.

We've seen already how fish weirs work with their hedges of woven willow on stone walls, and stake-nets work on the same principle, although much more refined by leading the fish into chambers that they are unable to escape from. The walls are made from netting carried upon the stakes that are set into the ground and are always sited so that they dry out at low water. A paidle-net was a miniature stake-net and often set to catch flounders.

Stake-nets were subjected to various restrictions and, in 1867, came the first instance of their rights being removed unless they could adhere to two requirements: they had to prove their gear was fished for one or more open seasons between 1861 and 1864, and there was a grant or charter of memorial usage. This was the first instance of confiscation without compensation of private or a heritable right to catch salmon. Certificates were issued to those that fulfilled the requirements and subsequently, anyone who wanted to fish in this way in the Solway Firth had to acquire a certificated site and erect gear that conformed to the requirements of that certificate. The last stake-net at Annan was the Snaab net, which was out on the sands to the west of the remains of the old railway viaduct, and its remains can still be seen to this day. Unfortunately, I didn't have the time to venture out onto the sands to view them.

John Warwick, who sent me the photos of this structure, told me how his father had a hut on the viaduct that was used as a base during fishing time. He had fished full-time during the summer months and in the spring and autumn he would work at putting in and taking out the stake-nets. John, who comes from generations of a fishing family, started going out fishing during summer holidays with his father, who was known as Slogger. He recalls how his father carried him over the 'leads' on his shoulders (too deep for his wellies). Initially he said he would just play and explore, enjoying the seaside environment, though as he got older he was given jobs such as helping carry the fish off and the delivery for fish orders. Trout were wrapped in rhubarb leaves to help keep them fresh and fish were carried in bass (straw) bags. John eventually started fishing with his own licence at the age of 18 at Loch and Dornoch in return for helping put in the stake-nets. A few years after that, he got his haaf-net licence at Annan and continues to fish, even though, as a career, he went into teaching. That enabled him to fish during the school summer holidays. Now retired, he has greater flexibility and only fishes occasionally. Despite that, he says he still enjoys haaf-netting just as much as he did when he started. Today he is chair of the Annan Royal Burgh Fishermen's Association.

'There's the pride in carrying on a local tradition,' he told me. 'The licence monies from the stake, poke and haaf-nets have, over the years, been the main source of income for the Annan Common Good Fund, which in turn has helped the wider community. Sadly, now it is only the haaf-nets that continue to fish and the Common Good Fund has had to restrict its support for local good causes.'

Poke-nets are said to have been unique to the Solway and of considerable antiquity. They were simple netting pockets about a yard square, mounted on poles driven into the seabed, in lines across the current, just like Adrian Sellick's nets in Bridgwater Bay or those shrimp nets at Weston-super-Mare. The difference is that the bottom corners of the nets had rings attached allowing the net to ride up the posts when there was a fish inside, the pocket collapsing as the tide ebbed. They were set in groups of four, this being a 'clout', and each fisherman was licensed for fifteen clouts, or sixty nets. They had been in use on both sides of the Solway until being made illegal in 1865 on the English side. In 1877, when evidence was being given to the Special Commission for Solway Fisheries, whose main task was to look at the legality of fixed engines (stake and poke-nets) on the Scottish side of the Solway, the town clerk stated that there were some 500 poke-net clouts both sides of the Solway railway viaduct. But, as we've already seen, those working the various fishing methods had to work together. Thus the 1848 (4th April) Stent Roll & Regulations set out rules for poke and haaf-net fishing regulations, by order of the Magistrates, John Sawyer, Provost:

Every Two-fourth and Three-sixth Stents to have a right to set Eight, and every One-fourth Stent, Four clout of Pocket Nets. If any dispute, in reference to fishing-ground for pocket nets, shall arise amongst the fishers, the Rig shall be divided, and each fisher shall take the ground assigned to him by lot. When any dispute shall arise among the Half-Net fishers, the Water Bailiffs are to give notice to the proper quarter, when all grievances will be redressed. The Mells to be cast for Halfing every flood and ebb tide.

Another extract from the 1877 Solway Commission suggests that up until 1870 the town allowed poke-net fishermen to fish rent-free. Presumably this applied to haaf-net fishermen also:

For a long series of years, a period of more than 150 years at least prior to 1870, the Town Council of Annan permitted the proprietors and inhabitants within the burgh to exercise this mode of fishing rent free, but in about 1870, the council took the fishings into their own hands, and have since regulated and let them yearly by granting licenses or certificates to the fishermen at a fixed rent for each 'Clout' of nets. A 'clout' consists of about four yards of netting in a straight line, fastened to poles about 4½ feet in height, fixed in the sands. The length of the line of 'clouts' varies according to the number of 'clouts' each fisherman may be permitted by license to use.

By 1962, the number of poke-net licences issued at Annan was thirty-four, and these were the only ones remaining on the Solway. Numbers slowly declined until they were removed in totality in 2015. Stake-nets, along with all set-nets, followed suit on the Scottish side about the same time, even though they had already been banned on the English side for a number of years, until only, as we saw in the previous chapter, the haaf-netters remained, and only twenty-seven of them at that. For how long, no one at present knows. We can only hope.

The one common thing both stake- and poke-nets have is the amount of work to set and maintain them throughout the season. At the beginning, posts have to be driven into the seabed, in later days with the help of a tractor, and the nets fixed on. Then during the fishing season, with weed a constant irritant, the posts have to be cleaned weekly, if not daily. Posts break and need replacing, nets get torn and need mending. It's extremely labour-intensive, and the returns perhaps not as forthcoming over the last fifty years as the fifty before them. Then, when the season is over, all the netting and ropes need removing and bringing ashore. Posts can remain, which today betray their existence around these shores as many still stand; epitaphs, if you will, to these once vibrant vernacular fisheries. Time, then, to move on.

26

THE VARIOUS YAIRS
OF KIRKCUDBRIGHT

From Annan, we ventured to Dumfries and followed the River Nith along its eastern bank, past Kingholm Quay, part of the port of Dumfries and built in 1747, to Kelton, one of the places where the Nith haaf-netters work from. Here the river was running strong. I had read a description that likened the Nith haaf-net to a huge billowing bra, the salmon and other fish being enticed into these cups! The same article informed that they used shaved cedarwood for the beam, which was supported by three ash rungs.

Further south, the largest haaf-netting station on the Nith belongs to the Caerlaverock estate. They control some 4 miles of the riparian fishing rights on their side of the river. Permission to haaf was primarily given to the estate workers and folk living in the parish villages of Glencaple, Bankend and Shearington. However, according to the Nith Haaf Netters Association, many people hereabouts don't even know haaf-netting exists. In 2018, there were thirty-eight licences taken up on the river but maybe half didn't venture out. The association's chair, Tom Brown, is trying to get local folk to take up their case with Marine Scotland and politicians. Anyone, in fact, who will support similar arguments heard in Annan. In 2017, the River Nith haafers were allowed two salmon for personal consumption after the river was re-categorised as '2', although they later admitted to making an error and changed this to only the area north of Airds Point, which meant that the New Abbey side on the west bank of the estuary, and some of Caerlaverock, couldn't catch salmon!

Without young blood coming in, the traditions will disappear and we already know that there aren't the young ones coming along. The average age of haaf-netters is over 60, while some have been doing it, like John Warwick, since they were at school.

The Nith flows into the Solway at Carse Sands and between there and the River Dee, along what is called the East Stewartry Coast, there are several stake-nets including the Almorness, Torr Point and Balcony Fisheries. Then the first crossing point of the renowned salmon river, the Dee, is at Kirkcudbright, to where we drove, some 30 miles south-west of Dumfries.

Like stake-nets, cruives and yairs are things of the past. Whereas cruives were basic traps set as fences across rivers with wooden gratings that hold large fish back while letting small ones escape, yairs – or stage-nets – are much more varied in design. They were generally employed in estuaries and lower rivers. But before we go into detail about those in Kirkcudbright, I will mention floating-nets first. Variously called a cairn-, pot-, or croy-net, these were short lengths of net up to 40ft in length that were attached at one end to cairn or croy (a natural or man-made protrusion or jetty projecting into a river and used to manage river fisheries, providing an obstacle to slow down current, a shelter for fish, a funnel to net them, and a platform to cast from), and allowed to stream downstream. Kept up by floats, salmon that passed by would often become entrapped by their gills. A hang-net was set across a stream at slack water so that the salmon swam into it. The stell-, or cleek-net, was a form of sweep-nets that were worked by a coble.

Now I've mentioned cobles before without any explanation as to what one is. Put bluntly, it's a type of open boat developed purely for salmon fishing in Scotland and is not to be confused with a coble from south of the border in Northumberland, Durham and Yorkshire, even though many of them fish for salmon. There are similarities but we'll ignore them for now, simply adding that the English coble is regarded as having something quite quintessentially English about it. Mind you, the pronunciation is different too, with the Geordies calling them to rhyme with 'noble' whereas to the south, as well as north of the border, they prefer the 'cobble' sounding.

The Scots coble is a flat-bottomed vessel, coming in various sizes up to about 30ft, wide in the beam and with an upturned stem to enable it to work off the beaches. Before the adoption of engines, they were mostly oared, and subsequently some had inboard engines and others used outboards through a stern well. They were basically pretty stable platforms to work nets over the side in relatively calm waters. A coble was probably used to carry a fisherman over to his yair, although the short film I viewed, highlighting one of the Kirkcudbright yairs, used a small rowing boat.

A view of a yair, showing the hedge either side that led the fish into the gap.

The thing about yairs is that they seem to me to be something from a sort of nineteenth-century Heath-Robinson-ish past. Something that went out with the horse and cart. Part of antiquity, almost, but effective and popular because they were so good at what they were designed for. In other words, good at catching salmon. But first, at this point, we need to distinguish between two types: the basic weir-type yair, which is simply a fenced pool that drains with the ebb leaving fish stranded, just like the weirs discussed in an earlier chapter, or the more 'active' yair, where a fisherman stands guard at the door to catch any fish exiting. A sort of haaf-fisherman-meets-fish-weir thing. For that is what happens: the fisher sits atop this gap for many hours, with his pocket-net sitting in the water below him, awaiting that telltale tug at the feeling strings. A reminder of the various fishing stages I've seen up and down in parts of Europe – the Italian coast and the River Loire in France spring first to mind – where he or she patiently awaits. Perhaps even the so-called Chinese Fishing Nets in Kochi, Kerala, India, so loved by the tourists.

I found an old map of the River Dee at Kirkcudbright dated 1909 and it marks four yairs: Fish House Yair as being the downstream one closest to the estuary and, in order proceeding upstream towards the town, the Gibhill Yair, Bishopton Yair and Castle Sod Yair. Werner Kissling tells us that local yairs were introduced by the monks of Tongland Abbey, though this isn't the first time I've heard about local salmon fishing methods 'introduced by local monks'!

I'm pretty sure that the film I found on YouTube was of the latter yair as it was closest to the fishing quay. Remains of it can still be seen today at low water, although most of the posts were removed in 2001. Look them up: 'John Jamieson Fishes the Yair at Kirkcudbright part one' and 'part two', the latter being a bit more exciting as they catch a few fish. It was filmed in 1997.

However, there were obviously many more yairs on the river at one time. A 1837 plan shows seven on the stretch between Gibhill and the Boreland Burn on the north side of the town, and there were more further downstream. Today's Ordnance Survey map still shows two of them.

To describe their operation, I feel it is best to quote an account written in the early nineteenth century by an unknown writer, which I discovered in another article written by David Collin and first published in the *Galloway News* in 2012. As it is anonymous, then I reckon I am justified, like Mr Collin, in copying it. The only comment Collin makes was that he remembered the wings being clad in nets, not wattle, such as is seen in Jamieson's film. Here goes then:

> The construction of a yair is not complex, and is easily understood. It consists of two parts: the wings, or leaders and the trap or poke. The wings are intended to decoy or guide the salmon to the point at which the poke is placed.
>
> The wing of the yair is a wall, or hedge, consisting of a row of stakes, closely wattled and interwoven with the branches of trees or rice (brushwood), so that the passage of the salmon is precluded. The length of these barriers is from thirty to eighty yards, and their height at the place where they approach nearest to each other is, in general, about 12 feet.
>
> Of these barriers, the water-wing, or that nearest the middle of the river, is generally placed in the channel of the stream at low water, and in some instances it is situated altogether in the channel. This wing is also generally shorter than the land-wing, in consequence of the difficulty of fixing the stakes and rice (brushwood) in the strong current and deep water of the river.

These wings, which diverge towards one extremity, to a distance of from 30 to 60 yards, gradually converge towards the other extremity, to a distance varying from 12 to 15 feet; across which, and resting on the extreme top of the two wings, there is stretched a bridge, consisting of a rack or hurdle, closely wattled with small rods, which retains the wings or leaders firmly attached together, and serves at the same time as a seat for the fishermen, and a place whereon he may deposit the salmon when taken. This hurdle, or flake, as it is called, is about 6 feet broad. The space under the flake, and between the convergent extremities of the wings, is termed the eye of the yair.

A flood yair is that which has its wings, or horns, expanded down; an ebb yair is that which has its wings expanded up the river. Of these, the former, for reasons that will immediately appear, cannot be fished until after the tide has more than half ebbed.

The manner of fishing a yair is as follows. To a rod, of a length extending across the whole width of the eye of the yair, is affixed, along the whole of its extent, the lower edge of the opening of a poke-net. Six yards in depth from the mouth to the closed extremity. The two side edges of the orifice of this net are attached to hoops which embrace and run upon two posts, one placed on each side of the eye of the yair, close upon the extremities of the wings. To the middle of the upper edge of the poke-net a rope is fixed, which again is tied to the flake, and by its tension the mouth of the poke-net is kept extended across the eye of the yair, and the upper lip prevented from falling to the bottom. A long rod, or wand as it is termed, passes right across the centre of the mouth of the net, and is fixed at right angles into the rod at the bottom.

The net, when fully expanded, does not occupy the whole space of the eye of the yair, but forms a parallelogram, of which the upper side rises in the middle, in consequence of the tension of the rope; and although corresponding exactly to the width of the eye of the yair, it is only about 6 or 7 feet deep. When the fisherman has taken his seat on the flake, he forces down the net across the eye of the yair.

The run of the tide keeps the poke-net fully extended. To its upper side, besides the rope fastened to the flake, there are attached four or five strings which reach from the net to the fisherman, who holds them in his hand. These warning-strings enable the fisherman to ascertain the moment when a fish has entered the net, and he instantly pulls up the wand and lays it across the flake, thereby closing the mouth of the net, which he then hauls up, kills and takes out the fish and replaces the apparatus as soon as possible across the eye of the yair.

As the tide continues to flow, the water rises above the poke-net; the fisherman then pulls up the wand a little, unties the rope that secured the net to the flake, and shortens it so as to keep the upper edge of the poke-net nearly on a level with the surface of the water. This is rendered necessary, because if the apparatus were to remain close to the bottom of the river, (as is the case when the fisherman begins to fish a flood-yair), the pressure of the water would be so great that there would be difficulty in hauling up the net; and it is believed, moreover, that the fish generally swim nearer to the surface than to the bottom.

It thus appears, that a yair is a machine calculated only to conduct such fish through the eye of the yair as may in their passage up or down the river chance to get between its wings. That, of the fish thus guided, those only are caught which happen to pass at the very time when the fisherman has his apparatus arranged across the eye of the yair. That during the rise of the tide, the space between the bottom of the poke-net and the bottom of the river, by which a free passage is afforded to the fish, is always increasing. That in flood-yairs, the whole time after half-flood, and in ebb-yairs till the tide has more than half fallen, the eye of the yair is left entirely free; for this obvious reason, that by the time the tide in the Dee is half flood, it has risen so high as to cover the flakes of all the yairs whereon the fishermen sit; and they cannot resume their places until the tide has left the flakes bare, which does not happen till after the tide has ebbed more than one half.

The New Statistical Account for Scotland (NSA) of 1845 mentions bag-nets below the bar, at the Ross, which appear to have been only in their second year of operation but 'we believe the fishing has proved remunerative'. Bag-nets were also in use off the entrance to Brighouse Bay, just around the corner from the Ross. Furthermore, the report mentions two 'common stake-net' fisheries at Kirkandrews and Knockbrex, both of which were abandoned by that time.

Going back even further, the Old Statistical Account for Scotland, of 1763 (Vol. 9) mentions another unusual method of taking salmon. At Tongland, there's a method called shoulder-netting, which was practised at night-time:

There is a small net fixed to a semicircular bow of iron, and this is fixed to a pole of about 18 feet in length. The fisherman ties a small piece of bended wood, with a groove in it, upon his left shoulder, for the pole to slide in. Thus equipped, he takes his station in the night, upon a rock at the side of the pool where he knows salmon lie, ands throws his net straight before him into the water, into which it sinks, and draws it straight to him on the bottom, sliding the pole upon there groove of wood upon his left shoulder, and when he has it near him, he gives the net a quick turn over by the pole, and brings out a fish; and there is another man standing close by with a club, ready to take hold of the fish and kill them, and take them to a safe place. In certain places of the river, great numbers of fish are taken in the night time by this mode of fishing. For the purpose there are two shoulder-net men, and one to kill, generally employed through the fishing season.

One reports suggests that 315 salmon were caught this way in one afternoon in July 1836. It simply serves to illustrate the prodigious amount of salmon that was being fished in the nineteenth century.

27

GALLOWAY ADVENTURES

Driving westwards from Kirkcudbright, we reached the main A75 trunk road and passed over the Water of Fleet where the Ordnance Survey map notes the stake-nets just off the island of Cat Craig. We arrived at the shores of Wigtown Bay, where the road almost touches the shore and, unexpectedly, spotted a sign for a 'Smokehouse deli and Snack Shack'. I quickly slammed on the brakes and swung into the car park. Inside, I was amazed to see 'wild smoked salmon' and, after enquiring, found that this comes from the upper reaches of the River Cree, at Bargrennan. Here, because the river has been graded as a '2' in terms of health, the owner, a Vincent Marr, who advertises himself as being the youngest son of the Marr family, is licensed to catch some wild salmon. The company sells 'Marrbury Wild Smoked Salmon', of which I purchased a small packet, as being that 'which HM Queen Elizabeth II had served at her banquet to welcome the world's Heads of State at the G8 Summit, Gleneagles, Scotland, July 2005!' Obviously this man has his contacts. I was informed that he fished a short season with a net and coble in the narrow river.

The Galloway Smokehouse is hardly a mile along the road. I'd been heading for this spot as I knew that out on the mudflats at low tide, the remains of the two Kilbride fisheries can still be seen – the Upper and Lower fishery. Inside I was introduced to the smokehouse owner, Allan Watson, who was extremely knowledgeable on the matters of stake-nets. He pointed out the Upper fishery, which was showing above the water and which, later, Ana and I ventured out to photograph as the sun was dipping below the hills across the short stretch of water. There were two more, he told me, just up the road opposite Creetown. All four were worked until a decade or so ago, these two by the Castle Cary estate and the two at Kilbride by one individual. However, he added, there were two more over the other side at Innerwell

(I'd already spotted them on the Ordnance Survey map). When the salmon came into the Wigtown Bay, they tended to wait as they gained their sense of direction. It is, after all, a long swim from Greenland. Some then swam into the River Cree and the four eastern side nets tended to catch these. Other salmon decided they were at the wrong place, so swam anticlockwise to run into the nets at Innerwell, which is a small bay just to the north of Garlieston. I read somewhere later that the Innerwell Port Salmon Fishery paid a rent of £200 in 1845, which shows that these are, indeed, ancient fisheries, and extremely profitable!

Allan was full of encouragement for my quest in documenting the last vestiges of the salmon fishery. We chatted a while and it was he who suggested I go down onto the mud to view the Upper Kilbride Fishery. As I glooped through the mud, it was a show of the work that went into these structures that most of the poles were still standing, even though it was at least eight years since this one, the last in use, had been worked. The poles stood about 8 to 10ft high, and I noted that, on the grass above where we walked onto the beach, sat some poles that were obviously once meant for the fishery. They lay in the grass by the shed that the fisherman had kept his tractor in – yes, they all used this mode of transport in the latter years – and measured about 18ft, which meant that they were at least 8ft into the mud, each one supported by a rope either side, attached to a peg driven into the mud. The first fence was some 170yd long, with the whole structure probably 300yd out into the bay.

Having trudged back to dry land, and driving into Newton Stewart to cross over the old town bridge, we decided to park up and take a walk along the riverside. Re-crossing the river over a newish bridge, I noted that it had been named the Sparling Bridge, and was a single-span pedestrian structure built in 1998, so called because the fish still breed in the estuary. Just north of the stone road bridge are the remains of the old wooden bridge and an old ferry, while it is said that the river was easily fordable in many places. Not good for salmon, I'm guessing, though anglers still fish the river and stake-nets used to be set in and around the estuary, such as those at Kilbride.

We drove onto Portpatrick, choosing to miss Port Logan with its infamous Fish Pond, which I was told was closed. Further south, I read that in the 1800s attempts were made to catch salmon with a draw-net off Monreith, but it was unsuccessful due to the rocky nature of the coast.

Gazing out over the sea from the cliff at Portpatrick, Northern Ireland seems but a stone's throw away even if it is 21 miles across! So often I'd been here at a much younger age, into the radio station with my dad when we'd sailed this way, and so often cloud obscured everything. Today, with binoculars, I can see the lighthouses of both Donaghadee and the Copeland Islands off Belfast Loch. This takes my thoughts back to 2007 and the circumnavigation of Ireland that resulted in my book *Working the Irish Coast*. I remember Portbradden, a tiny harbour with a few houses and a couple of stores on the west side of White Park Bay much further up the Antrim coast, away to the right around the corner, so to speak, in my vista. The Irish *Bradan* translates to 'salmon', thus telling us that this was originally a 'salmon port'. Then it was nothing but quiet.

Carrick-a-Rede was a place I'd heard and read about and I'd been determined to see for myself. This part of the sea was a particularly rich spot for salmon as they swam around the volcanic headland on their migration. This accounted for the high number of fishermen in the Ballintoy parish, which included both Carrick-a-Rede and nearby Larrybane. Carrick-a-Rede literally means 'the rock in the road', which refers to this obstructing the salmon's migration as they seek out the Bann and Bush rivers. The rock itself is an island, separated from the mainland by a vertical gap of maybe 60ft. Local folk have been fishing this spot by boat for generations, at least from 1620 and probably before. In 1755, the first rope bridge was erected across to the island, about 80ft above the sea, so that one end of the seine-net could be attached to the rock. When the bag-net was introduced to Ireland from Scotland in the mid-1800s, the fishermen here adopted this method, building themselves a net store on the island. Today, though fishing stopped in 2002, the rope bridge remains and crossing it is popular with visitors. On average some 200,000 visit each year, though not all cross over. Many come just for the bridge and have no knowledge, or a desire for any knowledge, of the salmon fishery, according to the people collecting the admission fees, and the site is the UK's second most popular attraction. Were they taking the piss, I thought? More popular than those London attractions such as Madame Tussauds or the London Eye, or the Giant's Causeway indeed. I doubted it. Perhaps they meant fishing attraction!

It had been late in the evening when I'd arrived in the car park, paid my £3 and the dog and I walked the mile or so along the headland, and a very pleasant walk it had been. The colder the August wind blew, the faster I'd walked. I wasn't the only person out, and when I did arrive at the bridge, I had to wait five minutes as the first group crossed and another group returned, the crossing operation controlled by the 'bridge-master'. Once across, I wandered about the island, gazing down at the hut and imagining how things would have been during the summer season when they had fished. In winter, the bridge was dismantled and stored away until the spring. There were plenty of birds flying around, the names of which I do not know, but I was aware that the island is frequently visited by ornithologists for its colourful bird life.

Not surprisingly because of its location, I'd read that a stream of visitors had been here over the ages, among them naturalist and clergyman Reverend Doctor William Hamilton, in 1784, who wrote:

> I went a short way off the beaten track to see a whimsical little fishing rock, connected to the main land by a very extraordinary flying bridge.

Mr and Mrs Hall came here in 1840 and described what they called the 'hanging bridge'. They weren't too impressed, for they wrote:

> The day on which we examined it was very stormy and we declined to cross it. One of our attendant guides ran over it with as much indifference as if he had been walking along a guarded balcony, scarcely condescending to place his hands upon the slender rope that answered the purpose of protector – the 'bridge' all the while swinging to and fro as the wind rushed about and under it.

In those days there was only a single rope handrail, unlike today. At its peak, though, they were catching 300 salmon a day and this slowly declined over the years as stocks depleted, largely, it is believed, because of the huge number being caught in drift-nets out at sea. There's an ice house above Carrick-a-Rede where ice was stored after cutting in winter, enabling the fish to be sent off to Belfast and England, and we'll learn more about these structures in a later chapter. Back at the car park – a bit drenched after a brief rain storm – it was back on the road.

Arthur Young, in his *A Tour in Ireland with general observations on the present State of the Kingdom 1776–79* (London, 1892), has a few notes about salmon fishing in this part of the country, which he says was considerable:

The fish are cured in puncheons with common salt, and then in tierces of 42 gallons each, 6 of which make a ton; and it sells at present 17*l* a ton, but never before more than 16*l* average for 10 years 14*l*. The rise of price is attributed to the American supply of the Mediterranean with fish being cut off.

Still, that's Ireland and we are straying off our track. But it is, of course, worth mentioning that huge amounts of salmon were once caught in many parts of Ireland, both in the rivers around the south and the west facing bays and lochs. Today, supplies are almost exclusively farmed. Enough of farming for now.

Continuing our journey up the coast from Portpatrick, the next stop was at Ballantrae, a small, neatly laid out village overlooking the Firth of Clyde. At one end the River Stinchar flows out, while a small quay affords some shelter in the tiny harbour. I remember that first time back in the 1990s when several boats were sitting on the grass above the harbour, and I still can recall their names: *Noreen*, BA63, *Margaret*, BA253, *Lyn Ann*, BA791 and the much smaller *Mary*. I was told last time that they were all salmon boats, though no other information was offered. They were certainly too big to work in the river, so presumably were fishing the estuary.

Margaret again in 2019 looking pretty forlorn.

More than twenty years later and *Margaret* was still there, though looking forlorn. *Noreen*, I was told by an elderly fisherman, had been burnt some years before and I couldn't help but notice the bonfire that had been built on the beach. This was the week before Bonfire Night, so it didn't take much imagination to see that this was probably how she had ended her days. The same informant did say that the last salmon fishing off the beach was in about 1991–92. They were 'ringing the salmon', he said, and landed some forty salmon. That's some catch in a beach seine and you can only wonder why they didn't continue.

The Stinchar was a thriving salmon river, while herrings were also in abundant supply. Maybe, for the first time, we see both fisheries meeting. According to the Old Statistical Account for Scotland, in 1791:

At the mouth of the river there is a substantial salmon fishery, which yields a rent of L.80 a year. The salmon are thought to be as good as any in Scotland, and fell upon the spot at 1½*d*. a lb. and it is but very lately that they were more than 1*d*. About 20 years ago there were great shoals of excellent herrings that came to these shores at the end of harvest, and beginning of winter …

By the time of the New Statistical account of 1845 we learn more:

Salmon go up the Stinchar, in great numbers, to spawn in October and November; and the young return to the sea again in April and the beginning of May. The salmon fishery at the mouth of the Stinchar, opens in the beginning of February, and closes in the middle of September, but the time both of opening and closing the river is considered too early. Though it opens in February, the tacksmen of the fishery never think of putting a net in the water till the 20th April, and even then not one fish in twenty is worth keeping, and at the time of closing, the fish are quite as good as at any previous time, save perhaps the end of July and the month of August. If the river were kept closed till the beginning of April, and not closed till the beginning of October, it would be a better arrangement, the Stinchar being considered a very late river. The rent of the fishery is L. 210 a-year. A market is found for the fish chiefly in Ayr and Kilmarnock.

The salmon boat Margaret, BA253, *at Ballantrae in about 1997 looking seaworthy.*

Although all commercial fishing has ceased, it is still the case that the River Stinchar is one of the most productive of the west coast salmon rivers. It's tidal up to the old road bridge, where the Bridge Pool sits happily between that and the new bridge. Here, I'm told, it's not unusual to spot up to 300 salmon lying quietly below the bridge. Today, though, this is the realm of the fly rod fisher and they are able to fish from various spots along the bank. Presumably it's a catch and release fishery now, though online statistics show a drastic decline in the numbers of fish being caught year on year.

We drive northwards, first edging along the Clyde, with Ailsa Craig standing clear off to port. At Ayr I think of the salmon river there but decide that it would be monotonous to write about it. There are many of these rivers and streams facing the Clyde that salmon entered but either all records of commercial fishing are gone, or they mirror that of the Stinchar.

The Clyde was once extremely rich in that other fish we are interested in: herring, and from harbours such as Girvan, the Maidens, Dunure and Ayr, and, to a lesser extent, harbours more to the north, sailing boats sailed out in search of these, the silver darlings in the nineteenth century, adapting to motor power in the first and second decades of the twentieth. There's an exciting story of determination and grit, sea adventures pitting man against the sea and its creatures, and which continued right into the 1970s, so that some of those fishermen are still with us. And these men used the ring-net with deadly accuracy in the post-war period and up to its demise when the herring ran out. There are names that stand out as being among the most successful of fishermen of that period.

However, it is not the Clyde that we are aiming towards, but Loch Fyne, that longest of Scottish lochs, and this entails two ferry trips, and two choices of routes at that: either the one across to Arran, a drive around this wonderful island followed by a short crossing from Lochranza to Kintyre or, the other way, over to Dunoon, over to the Cowal Peninsula and direct to Tarbert. The third alternative is a long drive over the Clyde on the Erskine Bridge, up Loch Lomond and over the quaintly named 'Rest and Be Thankful' to the upper tip of Loch Fyne and down to Tarbert, where we are heading. We had a vote and the children won with the first option that, as it turned out, was best in this instance.

The ferry leaves from Ardrossan. We were lucky as we'd been told it was full and were put on the reserve list. But it wasn't and we were aboard – they have a strange habit, Caledonian MacBrayne (Calmac), of saying that they are full but aren't. It happened on several occasions. Still, as we steamed out of Ardrossan harbour, I threw my eyes northward along the coast. Somewhere just beyond the harbour limits, and all the way up to Portencross, the visible headland about 8 miles distant, there were thirty-four possible fish traps among the rock and boulders, and another few around the corner. Although we hadn't had time to study them, there's an interesting paper written by Edward Patterson I have at home. He suggests that there are two 'supra-tidal ponds' where fish were kept alive after being trapped. However, a degree of caution is necessary and the author does stress that it should not be interpreted that they operated all at the same time, and that they may well illustrate man's efforts to obtain protein over a period of 3,000 years. Neither does he attempt to assume the type of fish that these traps might have caught, but you had to guess that salmon and herring could well have found their way thus far. We steamed towards the sinking sun.

28

YARNING WITH TWO TARBERT RING-NETTERS

As far as the west coast herring is concerned, and in terms of the gear used to fish it, then the advent of the ring-net was probably the most important innovation since the drift-net per se was invented. Other improvements such as the transition from hemp to cotton nets, and then to man-made fibres, vessel construction improvements, and the introduction of steam and the ensuing internal combustion engine, were all vitally important, but, in and around the Clyde, the move from the drift-net to the ring-net had far-reaching effects.

We had caught the ferry from Lochranza over to Claonaig, but before we left I had visualised the many Lochfyne skiffs that once sheltered in the bay. I have some lovely black and white photos from about 1912 that show skiffs dried out, their propellers emerging from the starboard stern quarter, for these were boats designed primarily for the ring-net, which was always shot and hauled on the port side. Shallow in the forefoot, easy to manoeuvre around a shoal and with a low freeboard for working over the side was an apt description of the skiffs. The motorised ringers weren't that different.

The ring-net developed from the practice of shooting a seine-net from the beach. In Tarbert, Loch Fyne, in the early nineteenth century, the fishermen, who previously had been fishing with their drift-nets in the loch, were experimenting, at first setting their beach seines and gradually transgressing into deeper water, using small trawl skiffs, so-called because once the net had been shot in a semicircle, it was dragged along for a short time before the ring was completed. It is this one fact that differentiates a ring-net from a purse-seine. And the thing to remember about the ring-net is that it was originally designed to fish close to the shore, not that it always was, of course!

The crew of the Campbeltown-built ring-net Lochfyne skiff Perseverance, CN152, *in about 1920.*

The introduction of the ring-net soon caused uproar among the traditional drift-net fishermen further up Loch Fyne, and they were joined by the fish curers, and then the authorities. They regarded it as destroying the shoals and their industry. Thus the mode of fishing that was regarded as 'trawling' was outlawed through government legislation. Fishermen, of course, continued and nets and boats were confiscated, hurting some of the Tarbert fishing families. Gunships were sent into the loch to police the fishing, which only managed to increase the resolve of the fishermen. The book *The Ring-Net Fishermen* by Angus Martin tells the full story of the violent birth of the method that, when legalised in 1867, was taken up by most of the drift-net fishermen once they realised its advantages.

There's a lovely description of the method of the procedure of ring-netting written by Dugald Mitchell in the very early years of the twentieth century in *Tarbert in Picture and Story*, which has recently been republished. He, too, calls it 'trawling' and talks about how the two boats work together, neighbouring each other. In his version, one boat holds one end of the net while the other, with the other end of the net aboard, sweeps it around a suspected shoal. He talks about 'pipes being smothered out' and 'the haliards [sic] whirred in the blocks' as they circled around. The 'crisp air was making me shiver in any case, but the thought of the unseen millions (perhaps) of fish that might be swimming in captivity in the shadowy waters of that circle – how near, how far away, you cannot tell – wrought me to an excitement.' The anticipation of what to find in a net is one all fishermen experience both from the hunting instinct and, today, from the wages that net will earn them!

But I wanted to learn more about what fishing with a ring-net meant and how they operated so I turned to Angus, who suggested I talked to two Tarbert fellows who'd been there. Thus, one wet February afternoon, I was sat comfortably in the front room of Willie Dickson, along with his colleague Archie Campbell, listening to the world that they worked in at the beginning of their fishing careers.

Archie started fishing in 1950 and Willie a couple of years later, and both started on ring-net boats. By then, the skiffs had disappeared, and the new breed of motor ringers were well advanced in their technicalities. Echo-meters had been invented.

'When we were young,' said Archie, who was the most talkative of the two, 'And we started first, it was the echo-meter was just starting, you know; it was not long after the war, and it was all new, during the day it was looking for gannets. If a gannet would strike, you should get a net around a gannet.'

Willie was the quieter one, soft spoken so that I had to move the recorder nearer to him. They'd both started working aboard the ringer *Village Belle II*, TT74, under the skipper Willie Jackson. The boat had been built on spec by William Weatherhead & Sons of Cockenzie in 1940, and when his earlier boat *Village Belle* was commandeered by the Admiralty during the early stages of the war, Willie Jackson bought this one from the yard and fished her until selling her in 1957.

'That was in the spring and early summer,' Archie continued. 'During the day, you know, when herring were feeding, and through the night you'd see them though the phosphorescence. We called it the "burnin". You would go forward of the boat with a mallet or hammer and bang on the stem of the boat, what we called "crepping", and the vibration moved the herring. The herring would be right below us, we were looking right down on them.'

Then Willie asked if I knew anything about the cartwheel. No, I answered, eager to hear more:

A big circle, herring inside and mackerel outside; we saw it first as a circle. The Dunure boys from Ayrshire told us what it was. Not a big market for mackerel; dinna want mackerel and herring together. You could see them in the burning, swimming around in a circle, from here to the main road [50yd], big. We were up in the north end of the harbour and we were seeing the cartwheels of mackerel. The Ayrshire men were in Tarbert that day and they told us, 'Put your net round it.' So they were still there the next night, making the way through the loch, so we shot and in one shot we got 250 baskets of mackerel and that was like about sixty cran and we thought that was enough for the market. So we got a good price. A strange market for mackerel. We used to be up in the Minches in spring and you'd land into Mallaig. If you went into Mallaig today with a few cran of mackerel you'd get big money for them. If you got rid of them tomorrow you'd have to dump them, no one wants them. First on the market and that was it. I canna mind what month it was, middle of the summer. April, May, June maybe. And we shot in this wee bowl of stuff and got the biggest mackerel I've ever seen in my life. Over 100 baskets. There were milt and roe in them. I've never seen a mackerel with milt and roe in since. They must have taken a wrong turning!

It seems the only big mackerel around these days are the ones that get into the fish farms and join the salmon with their feed. And coley and haddock too get in when they are small and feed voraciously and can't get out of the cages:

It was through the day we fished. The Minches in the spring, we used to go up about late April. The herring were feeding, you see. If you never got them during the daylight then you'd look for them at night. That was when you depended on the gannets. Feeding in the day.

I asked about porpoises:

Porpoises used to come up the Barra Sound, when the moon was in it. You'd hear them blowing and that. The old fellows would say, 'That's that, we're away home.' They'd chase the herring away. I was away at Canna, I was about 17, aboard the *Village Belle II*, with Archie, and we got five or six sharks in the net plus a whale. When they shot the ring-net it was open, you know, and they'd close the net. We had to wait to let them oot. They used to get a lot of damage with sharks. The basking sharks in the summer time. They'd feed along with the herring, fed on the plankton the same as the herring, you know, it was the same feeding they had. A lot of damage. And if you did get one you'd be as well to leave that net ashore because, whether they smelt it or not, it attracted them. Every time you shot that net you'd get a shark again, with the slime that come off them, the black slime that came off them. You cut that bit of the net out and renewed it. Nuisance they were.

We talked a little about how the ring-netting finished in the 1970s and how, when folk were saying it was finished, the only transom-sterned ringer *Prospector* was launched and how she neighboured with the cruiser-sterned *Stormdrift*, and that these boats earned so much they had to buy another boat just to offset the tax!

'All the Ayrshire boys were good fishermen,' they both said, almost in unison. 'Their boats were like yachts, all varnished like all the boats were back then.'

Willie said:

When I started there were still two or three boats with tillers working out of here, aye, just for the first two or three years when I started. Early '50s. The fishing started to deteriorate for a while then, and then it picked up again, came away again. There were a lot of boats, especially in Campbeltown, they all got grant boats after the war, you know, the grant and the loan off the government, and they had a big fleet in Campbeltown, lovely boats, but then the government took them off them, they couldna pay them, you know, the fishing fell away. They reclaimed the boats. They were lying at Grangemouth, the Navy took them over there, painted them grey, everything was painted grey! When we got the *Village Belle III*, she was one of the Campbeltown boats, she was all grey. We scrapped her when she came here, she was a Miller's boat.

We chatted a while about the various boats, and the conversation eventually turned to the Isle of Man, where *Village Belle II* was a fine and handy boat, always first up the river at Peel at the weekend to a good berth because she only had a 4ft 6in draft:

Most of the summer we saw the herring at the burning and if you saw a wee spot, a wee tick on the echo sounder, as well as seeing them in the water, you'd get a good shot. You see, when the meters first came out, they never trusted them. You'd see them in the meter but you had to be sure. You'd a feeling wire, you know, the wire with a lead at the end. You put the wire to see if you could feel them. Some of the men were better than others at seeing them in the water. Some of them were great eyes, when they took their eyes off the boat, of course. On a nice evening you'd stop the engine and just lie. You would feel them, out in the loch there, in Loch Fyne. You'd feel them jumping. You could hear them jumping, playing. We used to call it playing. At night, aye. You knew the difference between a herring jumping and a saithe jumping. You knew the difference, the different sound. The coley would jump, we'd call it a 'splooter' but a herring was a smart jump, a clean jump.

We were up Loch Fyne, at Inveraray, way up there one night with the Jacksons, and you couldn't see any herring at all. And the two boats came together and they hung out a light, with a 40 watt bulb, if you're lucky, and one of the men was lying up in the bow of the boat. And he says 'can you see them on your echo sounder as I'm hearing herring jumping here'. We shot and got 300 or 400 cran, a good fishing. They must have just been below the surface, came up to the light, as there wasna a mark on the echo sounder.

But I wanted to hear more about the actual process of ringing, putting the net out, so to speak:

Well, you just went slow through the spot of herring, right through as you are seeing it on the meter, through and when you come out the spot, over the side with the winkie, a light at the end of the net, and then you shot the bridal, that was a rope, and then the net. And you shot a half circle, right round the shoals.

I asked whether, by going through the shoal, that tended to disperse it or not:

At times it did, aye, but the thing is if you're at the shore, we were usually along the land, you've got to go the right way to the shore, the net to the port side and you see the herring. That's the only way to go, starboard side to the shore all the time. And if you're oot in the middle you had to shoot with the tide. If you never shot with the tide, the net would just pull away. Whatever way the tide is flowing you shoot your net so that your net's going out with the tide. If you shot against the tide, the tide will back your net away clear of herring. We had big nets and you'd hit the bottom at 60 fathoms. But they were too big, they would start to collapse with the tide, and you never got a sweet run. And that was all lifted by hand. So you shot the net and went round, and the other boat picked up the winkie, and then you towed for about five minutes and then you closed the net and the men from the other boat jumped aboard, your boat, and you lifted the net. And the other boat had a rope on you, keeping you oot of the net. Keeping the bow and stern oot of the net, you know. You'd a breast rope and you could manoeuvre it, the port side was the working side, that's the side you lifted the net, and the starboard side you'd a rope maybe up from the shoulder, a breast rope we called it. And the other boat had a rope on you, and he could keep you, if the net was coming under the propeller, or under the stern too much, if it was blowing, he'd keep it the way you wanted. Some of those engines going dead slow on tickover, especially the Kelvins, did that. Then you brailed the herring aboard. In the winter time up in the Minch, if you had a good ring of herring, maybe 200–300 cran, you had to get the net emptied quick because if the herring died they'd get very, very heavy, and you had a job to work them then. You had to empty the net as quick as possible. Two boats brailing. Sometimes four boats if nearby. If the herring got heavy in winter, you could lose them, they'd burst your net.

The heaviest was from now to the middle March, when the herring was spannie [spawning]. I've seen fifty baskets, you have a job with maybe twenty cran, you'd have a right good pull with them. They were that heavy when they were spannie. That was on the Ballantrae Banks, off Brown Head. It was only shallow, and it was only wee small nets we had, you know, but you did more damage as catch herring. You did a hell of a damage on the Banks, it was bad, bad ground, and you had to watch for the Girvan boats, they worked trammel nets, ground nets, and the stewies, we called them, the flags, the dhans, they would come up through your net, quite a job. It was a struggle the Banks. It's rough ground. All torn nets on the Banks.

But sometimes you got big herring mixed with wee ones. The big ones got the price so you had to sort them. You'd be in the hold all night sorting the small ones out.

I asked them how many different nets they carried aboard. 'Three,' was the quick answer:

One for the Ballantrae Banks, one for ringing by the shore and one for oot in the middle. For most of the year we carried two nets, you'd the net you were working in the stern and then you'd one on the backside, on the deck, on the opposite side from your working side. But when the pair trawling started, you'd a pair trawl in the stern and you'd a ring net in the stern. You didna know which one you were going to shoot. That was the beginning of the end. But it was a good job. But you knew your job, everyman had his own job to do. You knew your place. We used to take visitors out in the summer at home from here, holidaymakers who wanted a night out. And they were surprised at the way everything just worked automatic, the way everyone just went their own places, there wanna word spoken, everyone just knew what to do and that was it. Everyone to their own. Six crew, and some of them had seven. We had seven in the winter time.

Processing the catch: salted herring, barrelled and sealed, were sent all over while salmon were sent to market in wooden boxes.

They'd discharge near to where they were fishing. Ayr was good when fishing home waters as the rates were higher. When in the Minch it would be Mallaig or Oban. Oban was a better discharge rate but was eight to ten hours of steaming. Sometimes they'd stay in Oban and the old fellows would get a car home. Or maybe they'd then steam down to Crinan and get into the basin so that they'd all get home. Fishing the Isle of Man, they'd land in Peel, or Douglas or even Portpatrick sometime. Archie continued:

> Aye, when we went to the north, up in the Minch, on the wee *Village Belle*, sometimes we were loaded to the gunnels. You filled both boats, leaned over the side to clean yerself and then you set sail for Mallaig or Oban. Nobody thought much about it and all we had was compass and an alarm clock. No watches then. You set a time and away you went. They'd say look at that star if it was a clear night, follow that and that's how you steered. Sometimes it was a dirty night's crossing but we got there. Jumping from one boat to the other in bad weather, it was always risky but there were hardly any accidents. That's the way it was.

But I had to ask if they ate the herring. 'Did we eat them?' was the reply. Willie:

> Aye, we were browned off with them when we were young. When I was cooking on the same boat as him, I think I took over from him, there were six men in the crew, and there was four men came out every summer, ex-Tarbert men working away, but they'd always come aboard every night and I had to boil herring for about ten or eleven men. And was I fed up, a 17-year-old. And I had to make toast for the whole lot of them.

Archie:

> But boiled herring, that was it. We'd say, 'Not blooming herring again tonight.' And you boiled them in salt water, you just drew a bucket of water and boiled them in the salt water. One of our other pals, he was on a ringer too, and he was fed up of cooking herring and he got a couple of chops and gave them that. And there was another fellow, and older fellow, and he got some skate and he hung and it was hanging up on the mast for a week, and I said to him, 'When you going to eat it?' and he said, 'It's not ready yet!'

259

Ring-net skiffs at Tarbert. In the foreground is a sailing skiff, TT64, while pulled up outside Dickie's boatbuilding shed is a motorised canoe-sterned ringer, TT65.

But that was what they did. I remember my own grandfather take the skate home and wing them and leave them hanging on a clothes rope oot the back, and they'd be hanging there maybe a week before they'd eat them. No, I quite like skate, boiled and mixed up with an egg and fried. It is good that way.

A lot of the old men would eat any fish. In the Minches, you'd fish close to the cliffs and you'd flash your lights in and the *scarts*, the cormorants, would come flying out and they'd knock your bunnet off. You'd get them in the net too. Good to eat. We ate them and the guillemots, the dookers. You aren't allowed to now. But the people of Tarbert are called dookers. That's our name, they call us the dookers. My mother would cook them. With the guillemots you had to pluck them and singe them. But the scarts, the cormorants, you had to skin them. My mother skinned them and roasted them. They were all right. But Saturday afternoon that was our sport, one of the boats shooting the guillemots but they're protected now. I've seen a man shoot a big deer, aye, one time. Huge big thing. We tied a rope to him and lifted him as far as we could, brought him aboard. Massive thing. And there's an old man there, a great old fellow, you couldn't disagree with him, a decent man and he says, he disna swear either, and he says the thing's full of pox. Well we gave it to the local butcher and he cut it up. And I was fed up, we were all the same. Somebody says, 'Can we no have a plate of mince tonight!' We were fed up even of eating venison. All week, two boats.

It would have been easy to stay yarning with these two octogenarians for much longer, but the afternoon was beginning to turn to evening. A bit about the boats and cleaning of the hold, about keeping the fish fresh, ice sometimes in the summer, but boxing them up and how they'd keep the herring looking shiny with the help of jellyfish. The same jellyfish – the scouders – that stung you. And I remember one story of how once they were in the Outer Hebrides and the net was ashore. It had snowed that night, and then had thawed, and then froze again so that, by the morning, the net was frozen solid to the ground and they had to wait a week for it to defrost. More about the Ayrshire boats and the poor old ringer *Watchful*, once a well-known boat, ashore at Ayr, which was now rotting away, and many other stories!! But then the rain had nearly stopped and I offered to give Archie a lift back to where he stayed around the corner, to save Willie venturing out. But, as a parting shot, and after so much talk about eating habits, I had to ask them about salmon.

'Aye we'd poach a few of them when we could!' Almost in unison!

Enough said, because he didn't mean cook them in water!

29

CHASING GENEALOGY AROUND MULL

From Tarbert we'd driven up to Oban, having a quick stop at the Bridge over the Atlantic for a dog walk. Thanks to Calmac, we had managed to buy a last-minute ticket for the ferry to Mull even though they kept insisting once again that the bookings were over-subscribed. Thus, after another walk around a buzzing Oban, and a snack at the open-air seafood bar, we embarked on the ferry. On deck Mull appears at first as some giant monster, sleepy, then seemingly waking as we passed the tip of Kerrera. On through the gap between Lady Isle and the lighthouse on Lismore that I'd sailed through many a time in my old ring-net fishing boat *Perseverance*, built in Campbeltown in an earlier era (1912), even before Willie and Archie were fishing. Duart Point to port and then into Craignure, somehow selfishly thinking we'd not even had a taste of the Sound of Mull, with its memories ringing loud. Before long we were disembarking, with the T-junction onto the main circular road being surprisingly hectic, cars moving in all directions, some parking wherever they could and a general melee of folk wandering between the terminal and the boat. It was pretty chaotic and thankfully, after eventually turning left, I noted that a good majority of vehicles were going in the opposite direction. Quiet single-track roads are a joy to drive along as the scenery goes from drab, to wonderful, to dramatic.

For the moment we drove south, soon turning west, dodging the odd passing car, past the turning to Lochbuie, some 6 miles from Craignure, and on along the A849 towards Bunessan; through Glen More, the road incising through the green lowlands, the bulk of Ben More on our right. The kids wanted to catch a ferry from Fionnphort over to Iona, and Fionnphort was many miles west and the day was ticking away quickly.

We were over to Iona by three o'clock, the island I had not visited since arriving by my dad's boat in about 1970. A glorious late afternoon saw us wandering its narrow byways, an ice cream here, a pint there. It must have been almost seven by the time we were back on Mull and it was a long drive back the way we had come, through Bunessan and along the edge of Loch Scridain, until we turned left and headed over to the other side of the loch and up and over Glen Sellister and northward with stunning views over to the Treshnish Isles, Staffa and Ulva. We eventually discovered what must be one of the most simple campsites around at Killiechronan Farm: a field by the loch and very basic toilet block half a mile away! I was at my happiest barbecuing on the firewok by the van, the kids playing football on the grass, a glass of the red stuff to hand, as the blue hour began its descent!

Winding our way north again the next day, we decided against taking the ferry across to Ulva. The ferryman is summoned using a simple yet effective method: slide a small wooden door across in the white-painted signboard on the wall, uncovering a red-painted square. The ferryman will see this and come across to collect you. As soon as you see the ferryman on his way, you must slide the door back to cover up the red square. The slider above will tell you if the ferry is open or closed for business. Brilliant. We moved on to the wonderful Calgary Bay for a wander along the beach. Caliach Point, with the remains of its salmon station and ice house, only lies around the headland in a half-hour walk, but this was unknown at the time. I had been keener to find the pier at Croig and walk over to Port na Ba, which had been described to me as one of the most wonderful beaches in Scotland. And it was, and even the kids agreed. Ana even went for a paddle in the crystal clear, yet very cold, water! Then it was a return to the van, a drive through Dervaig and over the Mishnish to find the welcoming and perfectly timed Tobermory campsite.

Now, I must anticipate a few months and, briefly, leave the chronological order, for much of this chapter really belongs to a later date, the reason becoming clear in time. Months later, after describing our adventures to my friend Gez McLachlan, he later revealed that, while relating my story to his father, his father had informed him his great great grandfather had fished for salmon at Lochbuie. I asked for more details before delving a bit into that history myself.

I discovered, with some information from Gez's father, that his great great grandfather, George Rae, had fished there in 1899 as he was mentioned in the Fishery Board Report of that year. George had nine children with wife Alexandrina, who then died in 1898, and then another three with his second wife. Details of his time at Lochbuie are otherwise pretty scarce. He retired at some time to Dumfriesshire, where he died in 1928 at the age of 78. One of his daughters from his first marriage herself married a Dalziel and Alexandrina Dalziel (their daughter) then married a McLachlan, with one of their off-spring being Gez's dad. It would appear that George Rae was leasing the fishing from the Maclaines of Lochbuie who, many generations before, were akin to the Macleans of Duart, until two brothers fell out, and thus the spelling change of names.

George was fishing bag-nets that, until now, we haven't come across in any detail. Simply put, a bag-net consists of a long vertical wall of netting called a 'leader', often several hundred yards long, running at right angles to the shoreline, that is intended to interrupt the natural swim of the fish and direct them along it away from the shore and into a series of chambers or traps, the last of which they cannot escape from. The structure is anchored into deep water with the leader attached to the shore, and is then emptied by the fishermen on a daily weekday basis. It is set at the beginning of the fishing season in February (or sometimes they choose to wait some weeks) and remains until being hauled out at the end of the season in late August. It is a heavily regulated fishery in that the net must adhere to certain dimensions and cannot work at weekends when the leader is removed. This is called 'slapping the net'. The nets around the chamber also have to be changed, often on a weekly basis, which, all in all, makes this a labour-intensive fishery.

Although we are not sure of the date that George Rae left the fishing, documented evidence tells us that the decline in salmon catches was obvious by that time. In 1835, Scottish fishermen were sending 42,330 salmon into Billingsgate market in London each year. Between 1884 and 1893, the average annual amount was 23,749 salmon, decreasing to 17,160 between 1893 and 1894. In 1905, the amount was 14,368, a substantial decrease from seventy years before, though there could have been other reasons for the reduction.

Clan Chief Maclaine did fish himself and was instrumental in setting up the Lochbuie Marine University in 1887, with noble and rich directors from around the area, though the project seemed to not amount to very much and soon fizzled out. When Lochbuie Estate was sold to the Corbett family in 1920, the fishing was being run by the estate themselves.

Nets at Camas hanging out to dry. (Jane Griffiths)

The current owner, Jim Corbett, kindly phoned me one evening a few months later with information when I had emailed the estate. He wasn't sure of the details before 1939, but after the war he said that his father employed two men at the fishing. Then, by the 1960s, Kenny Gibson had the lease. His father, Alistair, owned the fishing boat *Glen Carradale*, which Jim said he'd sailed upon many times. I remember surveying the wreck of this fine ring-netter in Lochaline in 2000, at which time I, along with fellow students from the Scottish Institute of Maritime Studies at St Andrews University, had drawn up a wreck plan.

Subsequently, over the years since their purchase, the Corbetts leased the fishery to various people: among the names I heard mentioned were Peter Fennel, Eddie O'Loan from Northern Ireland, his son Danny, Peter Riley and the Gibsons again, fishing in the 1990s. Jim says that by the start of the new millennium, under pressure from anglers, they ceased to lease the fishing any longer although, presumably, the rights still exist.

Lochbuie has, over the years, had various bag-net fishing stations and I found four on the map: Rubha Dubh (Black Point), Rubha na Faoilinn (Gull Point), Port Ohirnie, and Port a' Ghlinne, Glen Libidil, the latter on the 'Back o' Laggan', as they say there.

Nevertheless, this initial information had given me the impetus to look further when previously I had simply not realised the presence of a serious commercial salmon fishery around Mull. And that quickly led me to

Allan McInnes and his seemingly clan of fishermen. For, between Lochbuie and Fionnphort, on the spit of land known as the Ross of Mull, there are three other salmon fishing stations that were fished up to the 1990s – Carsaig, Uisken and Camas – and Allan had a hand in two.

Of course, we also omitted to visit them, not knowing they existed at that time, although I have done on many occasions since thanks to the marvels of the internet and Google Earth; the very same internet that led me to discover a host of people unlocking some of the mysteries of salmon fishing around Mull.

First was Facebook, where I came across Bryan Gibson, who led me to Lochbuie and his father Kenny, as already described. I recall him recounting to me the words of his father: 'In the 1960s they'd be getting maybe fifty salmon a day but by the 1990s they'd be lucky with thirty fish a week. The increase in the seal population and the sweep-nets, even if they'd been banned by then, he reckoned were the causes.'

Then, thanks to Malcolm Burge, I was put onto a site called 'Mull: help and information', which, in turn, led into a mine of information from people who spent time at the salmon and others who had documented it over time. Over time, emails and telephone conversations, I learned how one man had come over to Mull in the 1920s and started what almost became a dynasty of bag-net fishermen around Mull and how incomers from south of the border had also built up successful fisheries employing local people.

The first of the latter was Peter Riley, who had come to Mull from Milton Keynes to study the geology of the island, according to Alastair Mackie, who worked for him for some ten fishing seasons in the 1970s and '80s. Alastair was fresh from school and Peter was fishing at Uisken. I spoke to Peter briefly on the telephone but he appeared reticent to say very much about his time at the fishery. When I asked him if he had any memories he just replied, 'None really. It was forty years ago and it's all in the past.' He couldn't get off the phone quick enough!

Alastair was much more forthcoming. Riley employed five boats with two men in each, working up to maybe a dozen bag-nets during the season, depending on the run of fish. These were mainly at four stations, he recalled, and nets were often doubled up using outriggers, while at Ardalanish Point he remembered three nets joined. The five boats were fibreglass with 40hp engines, unlike the cobles others used.

'He had no preconceptions, Peter Riley, and did it differently to the others who did it the way those before them had,' Alastair mentioned. 'For instance, we toggled up the net instead of tying it, which took a quarter of the time. He also employed local women who repaired his nets and packed the fish that mostly went to the Glasgow market by train.'

I asked Alastair his feelings on fishing. 'We laughed all the time. It was an enjoyable experience, wasn't that hard work. Long hours though!' That's something I'd heard a lot of. It was a peaceful occupation, engines switched off as they emptied and cleaned the net. The jellyfish were the worst, although the introduction of rubber gloves was wondrous for preventing stings. He also added that, when I asked about Peter Fennel, he remembered he owned Easdale Island at the time and set gill-nets at times. 'He would come over in his boat from Easdale, and sometimes stay in the bothy at Glenbyre, between Lochbuie and Carsaig. I think he had some fishing at Easdale too.'

A few months later I met Andy Walls, mostly through his rebuilding of a Morecambe Bay nobby at Lytham St Annes, who recalled buying a cottage from Peter Fennel. Remember me telling you about that evening at the Dock at Lytham? Well it was Andy who told me that Peter Fennel made his living by buying and selling old stamps:

> He bought Easdale Island and did up the old cottages that were almost wrecks there that the quarry folk had lived in. Must have been back in the '70s, maybe before. He set up some lads from the university in Glasgow to stay in a couple and they worked the fishery for him. Then he had some sort of processing sheds there too. I remember him trying to sell Easdale stamps as legal tender, which they weren't! I bought his last two cottages.

I never did ask Andy what he did with his cottages. It didn't seem important. I guessed the 'processing plant' must have simply been a shed to ice and box the fish. Still, it was worth mentioning.

Neil Cameron, who also worked for Peter Riley, described the fishing as 'enjoyable'. He worked for him and another fellow called Crispin Fryer for some twelve years, mostly in the period after Alastair had left, and reckoned Riley packed up a few years after he finished in 1988. Crispin Fryer and Peter Riley were married to sisters, I am told, thus making them brothers-in-law.

Neil did send me a sketch of how the bag-nets, with one rigged outward, worked. It seems the idea was to catch any fish that managed to escape the first net. The first net was set as usual with the rigger leader for the second net fixed to the cleek on the downtide side of the first net. Sometimes there were two outrigged nets, with the third similarly set, though, as Neil said, the leader might be shorter if the tide was strong. Outrigged nets such as these could only be set in water up to about 30m. Anything over that was too deep. Mind you, three nets in a line was a formidable barrier, which might explain Riley's success at the salmon.

Neil also told me about another bothy close to where they would set a bag-net at Shiaba, which is an old township settlement between Uisken and Carsaig, some 3 miles east of Uisken. Here, among some of the most fertile land in the area, approximately 130 people lived prior to 1847, when the land was cleared by the Duke of Argyll and the inhabitants forced elsewhere, some to inferior land, and others emigrating to America, Canada and Australia. However, he did add that the bothy wasn't used in his time and that they simply motored over from Uisken to service the net.

Allan McInnes had moved from Kerrera to Mull in 1923, probably around the time that George Rae was leaving in the opposite direction. He'd been fishing bag-nets with his brother on the small island of Kerrera, which is opposite Oban, but, with Allan being called away to fight in the First World War, his brother continued fishing on his own and had tragically drowned, caught in his own net alongside the pier, just to the south of Oban.

To begin with, Allan obtained the fishing lease for Loch Scridain, though we are not sure exactly where the fishing was. But it wasn't particularly good fishing in the loch, with a high seal population adding to the poor quantity of salmon. Allan and his wife Mary, along with their eight children, were staying at Feorlinn in Carsaig and so it was natural that he progressed to leasing the fishing there, working the bag-nets and helped by their children. However two daughters – Katie and Annie – along with eldest son Johnny, were the prime movers in aiding their dad. At other times he was helped by a retired man who lived near the pier at Carsaig with the heavy lifting. During the Second World War, there was an RAF camp nearby and some of the men there would help out and they also ran the fish over to Pennyghael on the main road to catch the bus to take the fish to the Oban ferry.

There are stories I heard from the family about how the girls joined Allan in the boat he used, with him on one oar and two of the girls on the other. This was probably Katie and Annie as they were the keenest of all the children. In time, some of the other children moved away from Mull. Life was, of course, hard, and in the winter Allan found other work as well as trapping rabbits as far along the coast as the Carsaig Arches, a good hike, carrying back a whole host of dead rabbits on a yoke over his shoulder. The pelts made good money and the family ate so much rabbit that, in later life, his daughters were heard to say that they'd never touch it again!

After the war, Major and Mrs Gordon of the Carsaig estate sold up to Dr and Mrs Gray in 1946 and, in about 1950, the Grays decided they wanted the lease back as they believed they could earn much more money by running the fishing themselves. According to Allan's daughter, Maimie:

> The Grays thought Dad was doing too well, and they took away the fishing; they were going to fish themselves. So he had to go someplace else to fish. Fishing was just what he couldn't do without. He was very, very good in the boat, and very, very knowledgeable and very independent.

This was considered pretty unreasonable to the family and, with a general clean sweep of the board of all who worked on the estate, the Grays became generally loathed within the local community. There was some amusement when they built (or rebuilt) a boathouse with tracks to drag the boat up that, on the first high tide, were swept away. On the other hand, the McInnes family were popular around Carsaig and, from various reports, it was always the case that Mary seemed to have had scones or pancakes on the griddle and a cup of tea ready for any visitors. It's no wonder the Grays were unpopular.

As a result of losing the lease, Allan obtained the rights to fish at Camas from the Iona Community, who leased the fishery from the Duke of Argyll's estate. Camas is a tiny old fishing station between Kintra and Bunessan on the north coast of the Ross of Mull, at the head of a bay called Camas Tuath, as marked on the Ordnance Survey map. The four pink granite cottages, built initially over 200 years ago to house the quarry workers from the stone quarry, were later used by fishermen, while today they are integral to an out-door centre run by the Iona Community. From quarrymen to fishermen, and then Borstal Boys to today's young folk, this place has seen a mix of many different people. If only we'd known of its existence then!

Allan was able to stay in the cottages there for the occasional night, though it is said that after their very first night there they had to return home to Carsaig to fetch blankets as the place was so damp and cold. There was no running water or facilities, except the springs from the days of quarrying that the workers had channelled to create pools. But it soon became a haven and various family members, nephews, nieces, the lot, would come and muck in. Douglas Canning noted that even if family members weren't involved in the fishing, they'd simply stay there in the summer as it was such an idyllic place.

Johnny McInnes married Peggy from Uisken and worked with his father at Camas, though sadly he died from cancer at a young age. His sisters Katie and Annie also helped, while Maimie had a job as a teacher in the primary school at Salen, where she lived in digs. There she met Tom Brunton, a joiner who had come from Lanarkshire to work as foreman for the house-building company of the Killiechronan Estate. They married in 1958 and some five years later, Maimie got the head teacher's job at the Creich primary school, taking the attached house with the job. Tom had left his job by then and had his own workshop, so a move to Creich was simple enough. About the same time, with Mary having died in 1962, widower Allan moved into the first new council house just built at Fionnphort, leaving Feorline after some forty years of residence. What a change it must have been, a transition from a small cottage to a modern council house on an estate!

Tom joined Allan with the Camas fishery, possibly after Johnny died, as did various other folk from around the area as helping hands at differing times. They generally fished three bag-nets, though, if the fishing was good, they might attach an outrigger to one net to expand its capabilities. 'To up the anti', as it was described to me! Fish, wrapped in leaves – sometimes with wild Iris leaves (Yellow Flag) – and brown paper and then packed into wooden boxes, was landed by the wall beside the Bunessan Pier and was put aboard the Bowman's bus for Craignure, but there were occasions in the past when the fish was taken out to the well-known pioneering passenger turbine steamer *King George V*, as she sailed back to Oban during one of her seasonal daily tours around Mull and out to Iona and Staffa. That would have been prior to her being withdrawn from service in 1974. At the time there were two bag-nets – the Boys net on the headland on the east side of the bay and Sword Point (Rubha a' Chlaidheimh) on the west side. Later on, with the introduction of the modern nets, there was another at Hen Point (Rubha nan Cearc), which is along the coast toward Iona.

Katie McInnes married Terry Turner in about 1960 and for a while both Katie and Terry helped at Camas. However, wanting a fishery of their own, they successfully bid for the Crown Estate rights for the fishing at Caliach, on the north-west tip of Mull, in the early 1960s. These had previously been leased to Alex Yule and fished by the Noble family from Tobermory. For the first year of fishing, Terry and Katie lived in the salmon bothy there until they bought a house in Dervaig. We will hear more of this branch of the family in a bit.

About this time, Maimie moved to Pennyghael primary when the one at Creich closed through lack of pupils, again as head teacher, along with a house for the job, and so Tom managed to get the lease for the fishery at Carsaig, bringing it back to the family after ten years away. That must have been about 1970, after the Grays had sold the estate to Chalmers Watson in 1962, who proceeded to split it up and sold part of it to Colonel and Mrs Purves a few years later. The date also fits with the time that Douglas Carson left art college to work with Tom not long after the move.

With most of the siblings leaving the nest, it was left to Annie to stay at home and look after her father. That was until she met Bertie MacRae, an ex-marine from the Falkland Islands. They married in 1970 and, before long, Bertie had taken over the Camas fishing from Allan, now an old man.

Generally, my informants here were Sandy Brunton and his sister Mary Corbett, Terry and Katie's daughter Sheena Walker, Jane Griffiths, Alastair Mackie and Douglas Canning, Tom's nephew.

Jane Griffiths had moved up to Mull in 1985 and within a couple of years was working at the petrol station at Ardfenaig, when, by chance, she met Bertie MacRae. He'd recently lost his crewman Willie Wood and suggested that Jane become his crewman aboard his boat and help him with his four bag-nets.

Bertie packed up the fishing five years after she joined him, which must have been the early 1990s, and Jane then took on the licence. She reckoned, fishing with Bertie, they were catching in the region of 500 salmon between the beginning of May and the closure on 26 August. When she started, the fish were being placed in ice in boxes and sent by train from Oban to London. By the time she finished three years later in the mid-1990s, 500 fish was more of a distant memory and she was selling her catch locally. Furthermore, with reduced catches and the payment of the £1,000 for the licence, it was becoming hard to justify continuing. Indeed, it was at that point she decided to write to the Duke of Argyll asking for a reduction in the cost of the licence. His rather impersonal reply was to sell the licence to the North Atlantic Salmon Conservation Organisation (NASCO), thus bringing her fishing to an abrupt end.

When asked why she liked it, she replied:

We just took our time. It was fabulous, working the summer season, long days and short nights, wonderful weather some of the time. Sometimes we'd stay in the old fishermen's cottages at Camas. Once Bertie packed up, I always found someone else to help me. Get the boat ready in April and we'd be fishing by the start of May. Slap the net at weekends. Then, in the winter, when everything was packed away, we'd do something else. Bertie was a joiner.

I didn't ask what she did during the dark and cold months, but I did ask her whether she ever got any comments as to her being a woman in what many believe is a man's world, and superstition often reminds other fishermen to be wary of women. She replied with a bit of a laugh: 'Bertie always reminded people that he'd learned from his wife, and that usually stopped them in their tracks!' Annie had indeed learned from her father Allan and had passed her skills onto her husband before letting him carry on without her.

'Bag-nets are environmental fishing,' she continued. 'They fish passively. We fished on the ebb so that the fish had a chance not to enter the chambers. On the flood the tide forced the leader against the door, effectively closing the net off. It's not like a continuous stream of fine netting across the tide that don't give the fish a chance to escape. That's what gill-netting or sweep-netting or whatever you choose to call it is like.'

When I asked if they'd ever used gill-nets, the answer was a very firm negative. 'They'd been banned,' she said, 'and we were honest fishermen!'

Mary Corbett remembers she was 17 years old when she helped her father Tom Brunton with his two bag-nets at Carsaig. Today she runs a market garden on the shores of Loch Don. Her mother Maimie never fished, she told me. She does, however, remember up to eight gill-nets slung out from the shore, though she couldn't swear whether it was at Lochbuie or Uisken, where Peter Riley, helped at times by Peter Fennel, worked several bag-nets.

The same was said by Sandy when he described the local anglers criticising the bag-net fishermen instead of the real villains, the deep-sea fishermen around Greenland, catching the salmon before they even had a chance to swim into the wild ocean. And the fish farming of course. 'You can easily tell an escapee,' Sandy, who fished with his dad intermittently yet continuously for two whole summers, told me. Sandy trained as a boatbuilder at McGruers, on the Clyde, before working away in Southampton for some years, and then returning to Mull.

Allan McInnes working on his nets at Camas. (Sandy Brunton)

'Their tail fins aren't developed after swimming around in a tank, and their gills are not properly formed. Also there's usually lots more lice as well as the flesh being different. You can always tell when you've got one in the net. Easy to spot.'

Echoing Mary's words, he said that the family were always against gill-netting as they reckoned they produced an inferior salmon. He continued:

It's the same for the gill-netted fish, which are often damaged and the flesh is bruised when it is cut. You can always tell. I remember how the three families ganged together to send fish to Barons, the smokehouse in London, who wanted prime bag-net salmon. They joined together when they bought netting and ropes, but were often happy to do their own things the rest of the time, working independently of each other for much of the time. But the London deal was a big deal and Barons said they wouldn't take gill-netted fish. They sent up polystyrene boxes, which we iced up the fish in – they also bought us each an ice-making machine – and the boxes went to Oban and by night train to Glasgow and onward to London. They were banded to stop anyone helping themselves to fish as often the weight of fish sent and marked on the tally in the wooden boxes, didn't always arrive at the same weight. On the other hand, Unkles of Glasgow, two brothers in the market there, reused the wooden boxes I don't know how many times.

Sheena Walker, when recounting what she remembered from when her parents Terry and Katie were fishing at Caliach, mentioned the same thing. The years of the 1960s and '70s were obviously good years. Abundant supplies of fish:

We maybe had up to twelve boxes, which we drove to Craignure for the ferry at 5 p.m., from where they went by road [BRS] to Unkles for the Glasgow fish market in the morning. All fish went the same way. For a while we sold to Barons in London, who introduced the polystyrene boxes, and we started using an ice machine. Previously I remember wrapping individual fish in iris leaves, then brown paper, and posting! But catches declined to the point where we sold only locally to hotels etc. and sold from the door salmon steaks etc. Hence the reason he couldn't afford to pay a man to work and so I worked the fishing for the three summers before we gave it up in about 1993. My mother also worked a lot whenever nobody else was available. I loved it on a nice day but I hated changing the net on a wet day. We also fished a few creels to earn a few pounds and I picked whelks.

We had just one net at the point, apart from one year when he tried a net below Treshnish Point but it did very little. I can remember my father spending a long time getting the net set right, changing it often and shooting seals to ensure as high a return as possible.

The first boat was brought from Camas, I think, and sank the first day but they got it up again and then my mother brought round her launch from Camas that my grandfather had bought with the money he earned as a stand-in for Roger Livesey in the cult film *I Know Where I'm Going*, some filmed in Carsaig. He gave it to my mother for her 21st birthday. A new boat was bought from Curries boatbuilder in Oban in the late 1960s and my grandfather then bought one the same. Much lighter and easier to manage, it was used from then on.

My father brought in lots of labour-saving devices. To save damming a river and washing the nets in it, he invested in a power washer and a tank.

He rigged up a device for pulling a rubber dinghy up the shore, which also pulled the dirty nets on a trailer. A wire rope was wrapped round the back wheel of a Ferguson tractor, which was jacked up on blocks, the wire attached to the trailer, so when in reverse the tractor pulled it up the shore!

He also set up a wee fish farm in the garage. He mixed the eggs with the milt in basins and then grew them up until they went in the swimming pool he had built years earlier in the garden. All went well until a toad got stuck in the water pipe and they all died – my mother was in tears! He did have further successes though.

It seems that he eventually sold the boat and gear to Neil MacLean, cousin of Alasdair MacLean who we shall meet in a minute and who fished at Caliach for a couple of years, and then Nick Turnbull, who had worked with Terry for a good few years, had a go but Sheena didn't think there were enough fish to make it worthwhile.

Sandy remembers the coble his father had at Carsaig. It was one of the north-east of England types called *The Mandy*, easily beached with the propeller in a tunnel. Tom pulled the engine to bits almost immediately, a two-cylinder Petter, he recalled. He also remembers being out in the coble when they'd come across Kenny Gibson in his. 'The old ones would have a good blather' over the gunwales.

He also mentioned the filming of *I Know Where I'm Going* and how Allan McInnes had been paid to travel down to the Denham film studios in London. Roger Livesey had been unable to travel to Scotland for the filming due to the fact that he was also performing in a West End play at the time (1945). Thus Allan did all the scenes in Scotland and the period in London was to enable him to be coached by Livesey. It's quite incredible that a fisherman from Mull ended up in what became a box office hit!

Douglas Canning, Tom Brunton's nephew, used to visit Mull from Lanarkshire for summer holidays in the 1950s while still at primary school. He says his first memories are of how segregated life was on the island: the leftovers from the feudal system whereby the large estates owned most of the land and, away from the main villages and town, the locals, the Muileachs, were forced to live in their rented crofts and pay homage to their lairds.

After completing a course at the Glasgow Art School, he then moved permanently to Mull and started fishing with Tom at Carsaig. He has fond memories of *The Mandy*, as well as the time he was fishing:

It was a balanced way of fishing, the best sort of fishing, with the lack of noise, and I always thought of it as a romantic way of life. I was really lucky to come in at a time when things were still being done the way they had been for centuries, passed down through generations. People used wooden boats, small Seagull outboards, cotton nets, hemp and manila ropes, cork floats, oiled cotton raincoats, sou'westers and rubber wellie-boots. This changed into the 1960s with the introduction of fibreglass boats, powerful Japanese outboards, lighter synthetic fibre nets, plastic floats, polypropylene and nylon ropes, Vinco27 plastic oilskins and synthetic boots.

Of course, when newcomers such as Peter Riley and Crispin Fryer started there was a bit of local controversy as they were using monofilament gill-nets ... although they later changed over to bag-netting. As well as a shift in the materials and fishing methods, this was also a time of changing culture. Fishing was being seen more as a business opportunity rather than an integrated lifestyle. The developing 'commodity culture' of the time was fuelling a different attitude to conservation. Traditionally the local lobster fishermen were very careful to return small lobsters and moved creels on regularly. More powerful boats, hydraulic pot haulers, electronic fish finders and navigation equipment, coupled with more creels, put pressure on stocks. People's expectations were changing and the arrival of shiny new pick-ups was becoming a symbol of success.

This echoed the words of others. In my mind this switch occurred during the Thatcher years with her interpretation of policies designed by right-wing groups and leaders such as Augustine Pinochet in Chile, after the ousting of Salvador Allende. A time when government policy moved towards the so-called free market that was led by demand. The slippery slope into globalisation and the ills of international trade deals, Chinese manufacturing and the gap between rich and poor accelerating at speed, some say!

Peter MacLean came from a family of blacksmiths who worked and lived in the smithy's cottage at Dervaig with his brothers, one of whom had inherited the work from his father Sandy, and him from his father. But after the war, the smithy business was down. The horse and cart was gradually being replaced by the motor vehicle. Thus the family acquired the fishing rights at Quinish Point, a bit east of Caliach, and Alasdair MacLean joined his uncle Peter at the salmon fishing when leaving school in 1954. They were fishing three bag-nets around the Point. Unlike at Camas, they started earlier in the season, which officially ran from 16 February to 26 August. They'd start in March to catch the spring run of salmon, which could be a good fishing. To start with he remembers that, because there was no bothy, they had to stay in a room at Mingary House, which meant a long haul of the fish back to where a vehicle could reach. Within a year they'd built a track for the van to reach the house, so that it could be run directly to the ferry and the Oban train south. Their best catch, it seems, was forty-eight boxes of salmon in one day, which Alasdair reckoned added to some 400 fish. They'd normally get 100 on a day. But, as the decades progressed, the catches declined until they, too, finally drew the curtain on the bag-net.

By 2000 the salmon fishing was over and those still working went creeling or trawling and, like Douglas Canning, eventually saved enough to buy their own boat. For Allan's grandchildren, Mary Corbett, Sandy Brunton and Sheena Williams, have their fond memories, as does Jane Griffiths, all of whom were quick to share them, as was Bryan Gibson, Alastair Mackie and Neil Cameron. Peter Riley retired to the lowlands, while Crispin Fryer went back south of the border and Peter Fennel sold Easdale and moved away.

Mull has managed to act as a time capsule in regard to salmon fishing, and no more so than in the family of Allan McInnes, and I realise just how lucky I am in coming across them by chance. Their memories have been soporific for me writing in the depths of Gloucestershire during the coronavirus pandemic and in some ways I'm thankful that this chapter came as a precursor to the end of the book, even though it's out of chronological order. The salmon may be all about gone, but they will return. Memories such as these, once gone, are gone forever.

30

FINDING FASCADALE

I came across Fascadale by chance, and I'm so glad that chance transpired. Mull was all about chances! We were in the bookshop in Tobermory, having left the campsite to purchase fish and chips from the mobile chip van by the clock tower, and being allowed a few moments of respite in such a shop after munching on the pier, excellent as they were, when I spotted a book called *The Leaper* by Michael Barrett. The cover was a beautiful photo of sunset, all amber and saffron and fandango pink, over an intense yellow-mottled sea, contrasting the dark landscape. And there, in the foreground, equally dark, was a Scottish coble, easily recognisable by its upturned nose. After studying the cover, the subtitle seemed somewhat superfluous: 'Adventures in a Commercial Salmon Fishing Boat'.

Needless to say, I bought the only copy the shop had. Sixteen quid, I think it was. A bargain! Self-published. Then, back at the campsite, I began to read how the author had spent three years fishing the various bag-nets they had at the small settlement of Fascadale, some miles north of Ardnamurchan. Which was indeed very convenient, as that was generally the direction in which we were heading.

The following day, by the time I'd reached chapter two, we took the ferry from Tobermory over to Kilchoan. Leaving the small ferry, I decided we'd visit the lighthouse at Ardnamurchan. Climbing to the top, having taken the tour, the fellow recounting the story of the light did mention a bit about fishing and that some salmon fishing had occurred at Kilchoan itself. Oh how I wished, after the event, that I'd talked to him more about the subject. But, on leaving the lighthouse, after the proverbial cup of tea and biscuit in the tiny cafe, we dropped into the village itself and parked up across the road from the wee shop-cum-post office. Outside, net poles were strung out with flags, betraying their one-time usage. Various boats jumbled around the edge of the car park, some seemingly consigned to the nettles. I wandered around while the kids stayed in the van, and thus photographed one particular boat that stood out among the crowd. She was a coble, like the one on the front of the book, named *Iolair* and was registered as OB226.

We drove north again, and then, using a map, found the turn off to this Fascadale, which was marked on the Ordnance Survey map. The single-track road seemed to follow a burn until we crossed a small bridge over it and ended up in a farmyard. I realised I'd missed the turning and retraced back, finding what was little more than a track. We bumped along, skirting the hillside and eventually reached what was presumably a car park by a gate. A wooden sign announced we'd arrived at the Ardnamurchan Estates properties of Fascadale House, Lodge and Cottage. Just to emphasise the fact, below was a council sign stressing the point in Gaelic: FASGADAL with FASCADALE below for those unsure. We parked up and entered through the gate.

Immediately, I felt as if we were almost imposters. It felt unwelcoming. The 'House' overlooked the bay, while the 'Cottage' was upon the headland, somewhat away from the beach. The 'Lodge' was further inland. It was obviously the domain of the holidaymaker, with properties rented out by the estate. The stone buildings were obviously the remnants of the fishing era. We wandered down to the beach, finding the odd sign of maritime life, such as an eye let into a rock. A seal popped up nearby, amusing the kids. Under the watchful eye of the holidaying inhabitants of the main house, we wandered a bit more, and looked into the stone build-ings. The 'House' was obviously where Roddy Macleod and his wife Bobbie, the lessees of the fishery at that time, lived whilst the 'Cottage' was new. Roddy, who came from

'Welcome to Fascadale', but it didn't seem very welcoming!

Skye, was described to me as 'having been born with a silver spoon in his mouth, but was a good fisherman'. It seems his brother was an MP, which probably explains that bit. But money had been spent on the house since the fishery was a working unit, that was for sure, judging by comparing it to the small photo in the book. The author Barrett had lived in the Bothy during his time there, sometimes with the other members of the crew, the remains of which I couldn't find. According to his descriptions, it was a pretty cold place, held together with little more than string, huge gaps under the door to let the icy blasts in, and the mice. Or was it rats!

I recall listening to a fellow who had once worked in the salmon fishery about his time. He had stayed in a bothy – I can't remember where it was but on the east coast – and had described it as 'a simple life'. I remember thinking it was not that simple when you might be up at 4 a.m. to slap the leader (removing it for the weekend closed season) or still working at 2 a.m. before even going to bed. Hours were long and home was a small room shared with maybe four others. The bothy at Portsoy had six bunks! There was no luxury in these places and when not fishing, time was spent making or repairing nets. Each fisherman usually carved his own needles and mesh gauges until plastic ones became the norm. By that time, though, the advent of car ownership meant fishermen could stay at home and drive to work. Wages were never that good and, in the 1840s, a weekly wage was 10s, which, by 1920 – the peak of the salmon fishing – it had risen to £2 a week. I'm not sure what Barrett earned but he seemed to leave each year happy.

The building with a corrugated rounded roof was presumably the ice store, while the shed at one end could have housed a boat. Oh how I should have read much more of the book before arriving. Anyway, I took photos and, once they'd got fed up of guessing where the seal would appear once again, we returned to the van and drove on. That night we found a campsite at Ardtoe, near Kentra Bay, and, after cooking tea for all, I settled down with a glass of the red stuff to read the book with renewed vigour.

By chapter four we were introduced to their boat and here I realised that the salmon coble *Iolair* I'd seen at Kilchoan was the actual coble. Wow, I thought, how much more had I missed? Should I have poked about the buildings much more and probably discovered old anchors and warps? I would have done if it hadn't been for the presence of watchful eyes in the house. Anyway, I later discovered that this coble was the first fibreglass one used in the commercial fishery in Scotland and had been built by

J. Sellar & Sons of Boddam, Peterhead. This company, established in 1871, were, as well as having a history of building wooden and then fibreglass cobles, also operators of salmon fishings between Macduff and Portessie on the Moray Firth.

The descriptions in the book are wonderful. Without going into too much detail, the author joined a fishing team of four. The first year, 1976, there was Roddy, Michael Barrett and two other fellows, Paul and Derek. The salmon season lasted from May through to the end of August, after which the fish stopped running. This particular bag-net fishery had been well established before the Macleods had taken it on. However, in the late 1970s, the numbers of salmon were fast depleting and the Macleods packed up sometime in the mid-1980s. They sold the fishery licence to a diving fellow, who tragically drowned, and subsequently the fishery was sold to the Ardnamurchan Estates and fishing ceased.

The house, the ice house and fishing shed, 2019. The bay is just off to the right.

So how did the fishery work? Well, they operated six bag-nets spread along the coast, two as far away as Ockle to the east, and one nearer Ardnamurchan. In Scotland bag-nets are regarded as 'fixed engines'. The others were near Fascadale. Mike, who started his new seasonal job in April, gives a good description of how the system went.

Before the first net can be set, much preparation work on this, and the boat, had to be undertaken. Because it was the first time he'd seen a net being set, he writes that 'every detail of that day is fixed in my memory'. Once all the gear had been loaded into the coble – which, by the way, was new that year, having been delivered from its builders on the East Coast – they motored out to a spot where the marks on shore were lined up. The hawser to the shore had to be fixed by taking the coble stern-first to the rocks, and once this was completed, the bag-net, which comprised three compartments – cleek, doubling and fish court – which were all separated by walls of netting. Once the coble motored away from the rock, with the hawser paying out, the various bits of the whole net went over the side, including buoys, until the end was reached and a thumping big anchor was dropped over the side. Then head and cleek poles had to be inserted to create rigidity, and the leader net, which could be as long as 90m, was anchored. In fact, it was all anchored firmly down, with the buoys and cork floats keeping it on the surface.

Nets had to be checked every few days, depending on the weather. Daily when possible but that's a hard task for all six. To fish the net, the coble with a full crew came alongside on the up-tide side and the net was partially collapsed by untying the head pole. The rudder was then lifted, followed by the side of the fish court, until they could hold the floor and roof together. The coble was then pulled over the net, forcing the fish into a bag, which was formed. The bottom and top ropes were lifted aboard the boat so that the lacing in the fish court could be undone. Once it was, the net was swung up over the gunwale and the fish spilled out into the bottom of the boat, after which they were despatched. The lacing was retied and the boat moved back and the rope securing the head pole to the bridal reattached. Sounds easy, eh? Well, with the boat moving in the waves, the wind blowing and weed and jellyfish, it was not. The leader had to be removed and replaced with frequency because of all the weed. All in all, fishing the bag-net was an arduous and thankless task, and in those days, with catches diminishing, not a very profitable one. In that first year that Michael Barrett fished at Fascadale, they landed 284 salmon, 864 grilse and seventy-four sea trout. Then, when the season was finished, they had to return to each net in turn, and haul it out of the water, into the boat, take it all ashore, wash each net and all the gear and pack it away. Once that was done, *Iolair* came out of the water. And hauling those bloody heavy anchors off the seabed was no mean task. Barrett returned the following year and then again in 1980.

I so wish I'd read the book before visiting Fascadale, but it wasn't just the book that led me to learn more. It appears that this fishery was one of the largest on the west coast, it having some 21 miles of coast line to operate over. Prior to Roddy – and the year he took it over is, as yet, unclear – a Mrs Powrie ran it after her husband died, keeping all the crew on. A Robert Powrie leased the fishing rights at Soay until the island was sold to Gavin Maxwell in about 1944, which gives some idea of the date Roddy took it over.

In the 1950s, there was a puffer (coal-fired boat) that sailed these waters and various flags were raised up posts at Fascadale sending messages to the puffer. 'Bag-net needed' or 'two leaders please' or 'salmon to pick up' meant a small boat was put over the side and rowed ashore, either delivering or picking up. Boxed salmon was run to the railway station in Oban for consigning south. However, this must have ceased by the 1970s as there is no mention in the book of a puffer. Salmon was taken to Kilchoan to catch the ferry.

The road to Fascadale seemed to lead to the end of the world, as do many other single-track routes in northern Scotland. Nevertheless, you'd be hard pressed to find another more peaceful. Equally so, maybe, but not more so! But just visiting Fascadale acted as a brief introduction to bag-netting, and now I was about to learn more from someone who, until recently, had spent most of his life fishing this way.

31

CUIL BAY

It's quite a drive from Fascadale, across the imposing landscape of Morvern, then across the Corran Ferry and south to Duror, but it was well worth it. Nevertheless, if I'd known, we would have detoured to Kilmalieu, where the fishing bothy still exists, while there was another bag-net fishery inshore of the Torran Rocks, a few miles to the north. But these snippets of salmon info, as was the fact that there were stake-nets at Inversanda Bay, were about to come my way. We'd actually passed Inversanda Bay, not realising what lay out there.

The headland of Rubha Mor sticks out into Loch Linnhe on its eastern flank, thus forming the pleasant sweep of Cuil Bay on its south-eastern side. Salmon swim along the coast and right bang into the bag-nets of Sandy MacLachlan, whose family have been fishing here for four generations. Sandy is a mine of information, having been fishing for most of his life. Here the fishery is owned by the local estate for, as elsewhere in Scotland, the salmon fishing rights are mostly held privately. Sandy was the tenant. His house was easy to find, being on the main road in Duror, and he'd kindly but a large red float atop his hedge as a sign. After introductions, we got comfortable inside and straight into the nitty gritty of salmon fishing:

Latterly we only had three bag-nets, but it was just my father and me. When I was younger there was my grandfather, my father and two of his brothers and a cousin of mine and myself. In the halcyon days in the 1960s there was lots of fish around. We'd get two or three thousand fish a year from this little station. But in 2015 they set a three year moratorium for the years [2016, 2017 and 2018], which we coped with. 2019 was going to be OK to fish, so we purchased new equipment that needs replacing each year and then the bombshell hit. They were going to extend the moratorium for another seven years, and then that became ten years. There were promises of some form of buyout, of diversification, but they came to nothing. We hardly got anything in the form of value.

He paused and a phone bleeped. For a moment I thought it was my own but then I remembered that mine was still set on its recording mode. It was Sandy's, which he ignored and continued:

Excuses after excuses, that's what we got. And it's always the stock. 'We want to study the stock,' they say. But there were never any proper stock assessments. Fish farming, now that's a real threat. Salmon escaping. The recent storm, Ciara was it. The one before the last one. At Colonsay some 67,000 fish escaped during the storm and it's quite likely this was an underestimate. These salmon have been bred as feeding machines as that's all they do in the pods. These cannibals swim up rivers and eat the small smolts and then we wonder why there's no stock. Recently the salmon farming went over the £1 billion in profit and it's a huge business, so we are fighting a losing battle. They say to the netsmen to bide their time but only, possibly, there's twenty-five to thirty netsmen left and only possibly six or seven of them under the age of 40. So they're looking at guys who are going to retire quite soon by natural causes.

But it was very enjoyable, an exciting life salmon fishing. Herring's exciting but salmon's exciting too. And everything else that goes with it. The different species you get. Tough life for seals and jellyfish, but a sharp, hard season on the west coast. We used to start in March, because for the Billingsgate market there was always festivals, usually horse racing and stuff like that, in the '60s and '70s, and very popular to have salmon in them days. We used to pump it down to Billingsgate. Glasgow market. When I was a youngster we had the railway line from Ballachulish and the track was quite close to the sea down here and we would weigh and box our salmon, then put them in the rowing boat and row near, then stretcher them and then carry the boxes over fences and to the train, that would stop. It was hard work and youngsters today wouldn't do it. But the salmon went to Glasgow market or changed at Glasgow Station and then to London the following morning. No ice or nothing those days. Bare wooden boxes. Aye it was a lovely life, but hard. We got some big fish too. You had a box for a single salmon, 40–50lb salmon. Then the season finished in August and we repaired and stored the nets. And come September you took on public works. Went to work in the Forestry Commission. I was good at machinery, so ended up driving excavators and things and there was always work. You could just go along and ask the foreman. So you filled in the winter with that, and then I got itchy in the spring again. Left that job and got the nets out.

We talked about how the bag-net was set. It seems that the metal pegs in the rock to secure the net have been around for, as Sandy put it, 'generations'. The size of the net has remained the same over this time for it's believed that this fishery might be 400 years old, although not the bag-net clearly, if that was 'invented' in the nineteenth century:

> So you set the net at right angles to the coast. And the salmon swims along and hits the leader that comes out from the shore. This is fixed to one of the pegs whilst the wings of the net are fixed with steel wires to two others. The end of the net is anchored to the seabed. So a salmon will swim in inches of water often with its dorsal just showing. And then he will follow the obstacle to the first door in the net. This is 6ft wide and into the tapered chamber, and then he finds the 3ft-wide door, what we call the 'doubling', and swims into there and then the final chamber door is only 6in wide and he'll swim in there. Each chamber has a floor and roof. The leader is about 15ft deep, deep enough though in thundery weather the fish go deeper. Length is about 30 fathoms, sometimes more.
>
> The distance from one corner where the cleek is attached to the wing to the other side was 16yd across the mouth but we sometimes reduced it to 14. It's about the angle between leader and the wing panel and if it's too blunt the fish turn away. You want them to go into the first door, of course.
>
> To get them out, the whole chamber is kept apart by a big pole with a rope round it, and you let go and you collapse the floor and the cover together and you work your way across using your hands gathering them till you get the wall of the net and you pin the top and bottom rope into the pins on the gunwales of the boat, so you've got this bag of wall netting with fish swimming back and forward in it, and you can pull it in and kill them, or you can lean over and smack them on the head. There's a lacing door there that you can untie.

I asked about the nets themselves:

Takes a whole week to make one of these nets. The mesh of the leader is about five and a quarter in diamond and the walls are four and three-eights and they taper down to four and a quarter in the final compartment. That's the legal minimum size and you're not allowed to go smaller than that. It was very heavily regulated, number of nets, length of nets, mesh size. Closed times, you had to have a net out of the sea six o'clock Friday night and were not allowed to put it back in until six o'clock Monday morning. Used to be Saturday lunchtime that it had to be slapped. Get the leader off as it won't work without it. Yes, very heavily regulated.

We drove down to the bothy overlooking the bay. The wind was blowing straight in and it was cold. Bloody freezing and you could easily imagine it being as cold in the early part of the season with the southwesterly blowing straight up Loch Linnhe from the Firth of Lorn, and the wild Atlantic before that. Hardy blokes they were, as most fishermen.

One of the nets would have been set right in front of this old building. Two others were away to the west, the farthest just over a mile away.

Sandy MacLachlan's bothy in the days when it was thatched with small skiffs drawn up. (Sandy MacLachlan)

The bothy itself, a stone building, was once thatched, so the photograph shows, and it is said to be 400 years old, which is probably the reason the fishery is regarded as being of a similar age. Once the building was twice the size that it is now, as an early photo shows. Before Sandy's time, when a crew of four were employed, the fishermen used to live here for the season. They lived somewhere else during the off season and they returned there for the fishing season. As described in the last chapter, this was no luxurious living but a place of hard work. Presumably the part of the bothy demolished was the ice house as there was no other structure nearby. One forlorn-looking lone net pole still stood among the rusting anchors, looking pointless, yet still proud in some way, like a monument to past times, just as epitaphs so often do.

Today there are no human inhabitants in the bothy, only sixteen bag-nets and more leaders, stored away at one side, with various poles for the nets, all appearing ready to go to sea. In the other wee room there was a French dresser that, judging by the way it leaned to the right, seemed about to collapse, although it was still largely intact, while Sandy assured me there was a stove somewhere hidden away behind the rest of the fishing gear. Various cork floats of varying sizes sat upon the top of the lopsided dresser. The old nets would have been kept afloat with cork from Portugal at one time, and wooden barrels; these nets all had plastic floats. Furthermore, although no one had lived here for a while, the building was dry below its corrugated sheeting roof. The same photo, mentioned above, shows three west coast skiffs and a barrel, which Sandy said was a barking tub:

> That was one of your jobs in the wintertime, making new poles. You asked the Forestry Commission if they could supply Sitka spruce or Norway spruce for our poles. We used to get them when the sap had come out of them near Christmas time, cut them down. We were allowed to go ourselves in later years, just take the boat trailer and the van up the forest, and the thinner ones were in the sea and the thicker ones for net poles. The net pole ones you left the bark on but the ones in the sea you had to take the bark off or they would just get waterlogged. You'd used to cut poles in the winter time, replace broken ones. Basking sharks would break them, they would hit the net and splinter them like match wood. The black slime of them would stick and the minute you got that, you'd take the net ashore as you'd not catch another salmon.

A view of the same bothy, which has been reduced in size. (Sandy MacLachlan)

Which echoed the words of Willie Dickson and Archie Campbell at Tarbert. Sandy continued when I asked him about changing the nets:

> If there's a lull in the weather in May and June time, and the weed grows on the seabed, and next thing then if a south-west breeze comes it, it breaks away off the seabed and it will literally sink the net. With cotton nets it was a nightmare. So the nets had to come ashore and be cleaned, others going out. Work, constant work.

Nets were generally cleaned using a home-made wire hoop attached to a wooden handle, a bit like a tennis racquet, which was used in a sweeping motion to clear the weed – called 'switching' in some parts.

So while Sandy waits ten years to salmon fish again – he's not holding his breath – he fishes 500 prawn creels, fifty on a train, which he sets around Loch Linnhe. In winter these days he makes the prawn creels rather than having to repair and renew his bag-net gear. He can buy them for £30 but he prefers to make them for £7. Prawns are picked up and taken to Eyemouth, and from there, flown to Spain, arriving the following day still alive. These are kept alive in individual cells, called tubing, immersed in refrigerated seawater, so that they can travel to the Continent. Surprisingly, I learnt that prawns account for 40 per cent of Scottish landings of fish, and creel-caught prawns fetch a much better price than trawled prawns, which are generally 'tailed' and sold for scampi.

And what was the bait, I asked.

Herring of course!

32

ON UP NORTH AND A LITTLE OF SALMON AND HERRING TOGETHER

Another long drive and a slog up the airy Bealach na Bà road, the twisting 11-mile route over the hills that reaches 2,053ft and seems to scare the living daylights out of some drivers. Personally, I've known much scarier routes over the Alps but I guess it's what you are used to. So then we arrived at Applecross, the intended terminus for this journey, though, by then, I had realised that this wasn't to be the case.

Applecross is actually five settlements in one: Milltown, Camustiel, Camusterrach, Culduie and Toscaig. Poking about the various inlets, I'd sort of assumed that Applecross was home to a substantial salmon fishing but on driving around the area, it soon became apparent that salmon did not figure here. The nearest, I was informed, were bag-nets over on the east side of Raasay.

Then I read a snippet that Lord Middlebrook refused to let anyone set nets and that he employed a full-time watcher to keep an eye out for poachers. Remember in those days that folk could be sentenced to death by hanging for poaching salmon although, in other parts, a poacher would have his ear nailed to a wooden stake for a specific time – called 'mugging' – as a punishment.

So maybe now was the time to discover a spot of history about the bag-net. According to what I read, it was a John Hector who set a floating-net in Nigg Bay, just south of Aberdeen, in about 1821, the first of its type. This seemed to have evolved from the stake-nets. Though it's impossible to state who rigged the first bag-net, between 1821 and 1828 this John Hector had twelve nets in the same bay. These nets consisted of only two chambers but, by 1841, it is said that nets had a third chamber added and that the design of the bag-net has hardly altered since those times. I suddenly remembered that Sandy at Cuil Bay had said that they sometimes reverted back to two chambers to reduce the jellyfish menace, and that these fish nearly as effectively as the three chambered ones.

The single track north took us to Fearnmore, where we met up with friends. Gazing out across the entrance to Loch Torridon, the sandy beach at Red Point on its north-west tip was clearly visible, and when, some hours later, we were walking over its sands, it was just as simple to see the red-roofed house our friends were renting. It was maybe half an hour's walk from the road, past Redpoint Farm and across the machair to the fishing station at the eastern end of the beach, one that is marked on the Ordnance Survey map. At the west end there's even the remains of a small slip with a winch, according to that map. The fishing station consisted of a bothy and a shed close to the beach, and a larger dwelling further inland. The only inhabitants these days were the Highland cows that wandered around, watching us. A cluster of heavy anchors – I counted more than a dozen – nestled in the grass just above the beach, abandoned from when they last fished. Judging from the number of anchors, they either set many bag-nets or were exposed enough to need more than one anchor.

Nets at Badentarbat.

The smallest of the buildings was a wooden building with a breeze block front wall facing the sea, corrugated iron on the roof and a concrete floor. Inside were leftover bits of equipment, a barrel, a fish basket, some bits of rope and some lumps of polystyrene. Nothing much of interest. The bothy, roofless and consisting of two rooms, held reels of steel warps and little else. Further inland was a larger building with two gable ends, which was possibly the house of the owner of the fishery, such as at Fascadale.

Set into the rock we found several iron pegs, one with a ring attached. Whether this was for a net or to moor the coble was unclear, though the latter was more likely, even if the slip was several hundred yards along the other end of the beach. Presumably they could launch from the slip and moor over by the buildings.

Reading in the museum at Gairloch, there's a reference to the salmon fishing belonging to Sir Kenneth Mackenzie under an old charter from the Crown and which was leased to Mr A.P. Hogarth of Aberdeen, who sent a manager each year to the principal fishing station at Poolewe. Fishing was primarily with bagnets. In the early part of the season the fish was boiled and packed in vinegar in kegs, each containing some 32lb of fish, while later on it was packed in ice and sent by fast cutter to Aberdeen and thence by rail to London. It was noted that 1883 was a bumper year, although there's no mention of numbers of fish.

We passed through Badachro on the way back, primarily to show the kids the remains of the fish trap, or *caraidh*, in the south-west corner of the bay. There are other remains around the back of Dry Island. I recall being here many years before and photographing the herring curer's stencil that he used to brand the barrels of cured herring, which I'd seen hanging on the white wall of a cottage just close to the pub there. I still use a blow-up version of that photo as part of my Kipperland exhibits.

From Gairloch we motored up to Ullapool, the town crouched on Loch Broom, which was founded by the British Fisheries Society in 1788 to encourage the herring fisheries, and designed by Thomas Telford. However, the salmon fishery probably pre-dated that.

Thanks to my good friend Mark Stockl, a boatbuilder from hereabouts, I was introduced to Ally and Donald Macleod, who had a croft on the west side of the loch. Although Ally had worked as a GP in Midlothian and Donald, one of his sons, was currently working for the Ministry of Defence, Ally had kept the croft working and had kept the fishing licence for many years, fishing each summer. We sat around their kitchen table chatting over coffee.

Taking the nets out at Badentarbat, the pier visible in the background.

Donald had taken over the licence from Ally in the last years of fishing and between the two of them, they got their history across. And they knew their stuff. The season lasted from February to 26 August, although they didn't fish much before the summer months.

'Yes, we were sweep-netting in the estuary,' Ally started off by saying when I'd told them what I'd been up to. 'There were no bag-nets within the estuary limits, the nearest being at Camas a Rubha, that's right out. We could only use sweep-nets in here. Usually the best place was across the loch from here, at Ardcharnich Bay down to the head of the loch and around the mouth of the river. The net we had, still got, and never used now, was a four-man net, a bit strong for two people.'

In his *The Salmon Rivers of Scotland* (1902), Augustus Grimble mentions that crews regularly fish the mouth of the Broom and Ullapool rivers in July, August and September. He also mentions that 'the Broom does not produce one-tenth or even one-twentieth of the fish it used to yield before the fore-shores of Loch Broom became studded with bag-nets'. Although he would be correct in the decline of the salmon stock, I think he has confused bag-nets with sweep-nets as I'm sure there were never any bag-nets studding the foreshore in the loch. I asked how big their sweep-net was.

'Sixty fathom,' says Ally. 'And the bag in the centre about six fathom deep. It was very heavy, heavy twine. We had a buoy in the centre. With a good bag of fish in the net, with the buoy to keep it afloat otherwise I've seen salmon just jumping over. It was quite a pull, quite a pull,' he says!

> As well as that there'd be another of at least sixty fathom of rope and where we set it usually, and we were working off the shore, legally we shouldn't have thrown anchors to the shore, there should have been a man standing on the shore and the net should be moving at all times. But if there's only two of us fishing, and both in the boat, for you had to be quick if you see them come in. You'd see the net as close to the shore as you can and then use the rope to pull the net back to the shore, not necessarily where your anchor is, but down a bit.

Donald got out a map and showed me where they fished:

> The church owns the netting rights on this side, down here, and Inverlael Estate owns the netting rights on the other side. We rented it off the estate and we finished in the '70s and then latterly we got it again for a couple of years, 1982–83. And then we got the 'Glebe' for six years from 1985 to 1990.

I asked to what the Glebe referred:

> The church there [Clachan] own some land, the netting rights and an area of land called the Glebe by the river, so we call it the Glebe. On the road just before the church there used to be an ice house, there's a wall by the road and you can see it. The ice would have come from the river and the minister still has the rights to fish on the river. Yes, we had the salmon rights until 1990 but the salmon numbers collapsed about the mid-'80s. We used to get some fish but the fish farm came in about 1985. The Wester Ross one.

We talked fish farm stuff for a while, though Donald wasn't sure that the offshore fishing for salmon didn't affect the loch. But the fish farm seemed to lose control of the sea lice problem for a while. This farm, the only privately owned one in Scotland I'm told, is one of the least intensively farmed ones around and seems to have a fairly good record. So whether it was the increase in seals, or birds feeding on smolts, or the spread of sea lice, or sea temperature and climate change in general, or pollution from farms or agriculture, or indeed cross-breeding by escapes from the farms that have led to this collapse, we do not know:

> In the early '80s when we were fishing you'd maybe get half a dozen fish. And on a good shot maybe about a dozen, so there were still a reasonable number of fish, but that pales into insignificance to what they were getting in the '60s and '70s.

Ally continued:

> Most of the biggest shoals were in late July, early August, particularly if it had been dry for weeks and the river was low as they couldn't get up. Every summer they'd be using up time and just cruising around waiting for the rain and then a good southwesterly gale and the rain for a few days and the river rose and then they'd be off. One fishing I remember was 11cwt. I regret it now but we used to regularly catch sixty to eighty sea trout.

I was shown a tally book Ally had kept for the 1968 season, showing all his catches and income. Fish that was sent away was motored down to Garve station for carriage to Inverness and beyond. While I was perusing the pages, Ally told a story about the bag-nets in Gruinard Bay:

> My brother-in-law told me, when they landed 100 salmon in one day on the bag-nets at Laide, they were given a bottle of whisky. So they had quite a large cupboard in the bothy and Charles opened it one day and there were rows and rows of whisky. Said why don't you drink it? Ach, we keep it to the end of the season, then we share it out!

There were bag-nets at both First Coast and Second Coast at Laide. Ally recalled sitting there watching six or seven seals feeding on salmon.

Donald told me a story about a fishing station on the Solway. On a map he pointed out a netting station on the north side of the Solway at Torduff Point, near Eastriggs:

> So basically you've got MoD land there, and the netting station called the Loch Fishery there which was privately owned. So, for years and

Salmon in the bottom of the coble. (Ken Lowndes)

years and years, the MoD had to employ a guard, for when the MoD acquired the site in about 1987, the netting station was isolated but they retained the right of access to get through the MoD. So apparently the MoD employed him for years and years to sit at the gate and his job was to let the fishermen in.

He added that there was an ice house there, as well as another at the Dornock Fishery a bit to the west.

There are two fish weirs at the head of the Loch Broom, one on each side, and we chatted about them a while. Again the Ordnance Survey maps show them, as does aerial photography. Remains can be seen at low tide. Both date back to the eighteenth century, the one on the Glebe side having been built by the great-grandfather of John MacLennan, according to a taped conversation with him in 1961 that can be found on the internet. This one belonged to the church and when the Rev Dr Ross was minister – he died in 1843 – 1,000 herring were once caught in it. John must have been in his eighties, and reckons it was last used in the middle part of the nineteenth century. An article on fish traps by Thomas D. Bathgate notes that he thought it was built specifically for salmon but MacLennan says he thought it was for herring. However, it was obviously effective for both! The walls were up to 5ft high, according to Bathgate, and there was a wooden sluice, which was opened on Saturday night so that it didn't fish on the Sunday. MacLennan correctly calls this weir a *cairidh*, or, as he says, the old word was *yair*, which he pronounced as 'gair'.

Across on the opposite shore, Mackenzie of Could owned the *cairidh* until he sold the Inverlael Estate. This Bathgate described as a 'very fine rectangular yair' with stone walls 4 to 5ft thick and up to 18in high. It was built in the eighteenth century and was still fishing in 1910.

There's a great legend about how herring came to Loch Broom. It seems that, whereas herring were to be found all over the cast, there were none in the loch. So some enterprising fisher wives sent a silver herring off to Lewis with their menfolk. These folk threw the silver herring over the side, fastened to a line, and towed it home. A huge shoal of herring followed the silver herring back into Loch Broom. As an after note, the author Roderick Mackenzie adds that the boat was painted black on one side and red on the other to cheat the Lewis witches!

We said our thanks and goodbyes to Ally and Donald, and Mark, with the promise to drop by again. Heading out of Ullapool north and then

...ushing the last pole down to secure the net after releasing the fish. (Ken Lowndes)

following a turn-off down a long single-track road, we came to the wonderfully named Achiltibuie, and to nearby Badentarbat where, close to the road but atop the beach, is a collection of anchors and slowly rotting salmon cobles, and various other mysterious artefacts. And, just below the bend in the road is the salmon bothy of the Badentarbat fishery. This used to be the salmon boiling house as bag-net fishing for salmon started here in the 1840s. Once boiled, it was placed in barrels with vinegar for export. Once ice was used, they would take this from the nearby lake and store it in the ice house, which had been where the shed across the road now stands.

I'd been in touch with Peter Muir, who owns the bothy and whose father Willie ran the fishery up to its demise in the 1990s. There were, in fact, two brothers who had inherited from their father, Bel Muir: Willie and Jimmy, who, according to Ally, had fallen out and thus split the bag-net fishing. Willie, the eldest, retained the best with one bag-net near the bothy and two more nets up at Achnahaird Bay, what they called the North Side, a couple of miles or so to the north. Jim had others either side of Badentarbat. Ally had also added that Willie was the first in this area to use polystyrene boxes to send his fish away to market. I recalled being here in 2011 when photographing the wrecked cobles.

Ken Lowndes lives in Polbain and when I was informed in Ullapool that he was one of the fellows who worked for Jimmy, I sought him out with a phone call. He simply said to come and I needed no persuasion.

At first, Ken was slightly reticent about talking and suggested I should view the two DVDs he had on the salmon fishery.

'You'll learn all you need on what is there in them,' he insisted, but I hesitated as I suspected they were the same ones I'd got at home. Nevertheless, I took them and promised to return them by post, which I eventually did. When I told him I was more interested in his take on the fishery, his experience working it, he opened up somewhat. We started by talking anchors and setting bag-nets out into the sea.

'Our cleeks were made out of old seine ropes, very strong and just lead through them. Once they were pretty worn they were just dumped, but they were still strong enough for our needs.'

He made a rough sketch on a piece of paper, talking as he drew.

'Basically here's the shore and here's your anchor here, and a rope coming up from it with buoys attached here.' He pronounced it as 'boo-eys', which threw me for a minute!

And then you'd have your bag-net attached here and these are the chambers and here are the wings and the cleeks attach to the corners of the wings to the shore. So when you are going to set your anchor you obviously need to get it in the right place. It's got to be the right distance away from the shore. You can't really adjust the cleeks as the leader is a set length so you have a rope of a set length which we called a measuring line – off the top of my head 40 fathoms long – which you'd take out from the shore and drop the anchor at the end, and then you'd go to the buoys here and attached your net. Basically you'd put the whole net out in a couple of hours.

The idea was to take your time and not rush it. The whole net is about 12ft deep. Always set at right-angles to the shore and to make sure the tide didn't turn on you. And generally you used to find it would fish on one side and not the other side. See, we had a net here, that's Old Dornie there (pointing at an open map), we had one here at Fox Point (Carn an t-Sean-Aoich) on Isle Ristol and this was just the way the fish would turn. So when the tide was in, this leader would bow and sometimes the whole net would bow. When the tide was really strong the whole net disappeared. So that these buoys, and the head pole keeping the net in shape, the whole lot went down. The first time I saw it on a sunny, flat calm day, we were rowing out to the island net and I'm looking and thinking it's not there. I looked back at Jim and of course he was just smiling there and when we got nearer the top of a pole was just showing above the water by a few inches. It would normally be a lot higher. We just stopped the boat and were sitting there, and suddenly the pole started coming up, and there was the net and the next pole came up and then we saw the head pole come up and all the buoys and everything. We went alongside the net because it was there, because the tide had just come off it.

We talked about how Ken had got here as, in 1975, he was in New Zealand. It seems he came back home after being wrecked on a 50ft ferro-cement boat in a storm:

When I got back to England I had a mate who was working in Ullapool. I was on my way to Iceland, thinking of going to the West Indies, I'd been there before, and back to New Zealand. Anyway I visited my mate and ended up at a ceilidh on the Friday night, fell in love with a girl. So she slowed me up for three weeks. And I never got away from the place. She

The salmon crew at Badentarbat, 1930s.

went back to university and I met Sandy Loots and Jim Muir and I went out with them. They had a hang-net, a monofilament, exactly the same as a gill-net, which they were then allowed to use. We used to work it for a while. You used to be able to sweep-net too in Polly Bay, the next bay up going north from Achnahaird Bay. This was all legal but that one had stopped by 1975. They'd have someone on the cliff spotting the shoals. Willie used a sweep-net there sometimes, although I'm not sure what happened because the fishing there was owned by the estate and I think there was a hatchery there for many years.

I asked him how Willie and Jim had split the area up between them:

Jimmy had the whole area from below Achduart right up the coast, not Badentarbat, straight through here under Polbain, Fox Point and right round the Rhu, but not that north coast. But these weren't as productive as Badentarbat and these on the north as well as at Reiff. So Jim and I were just the two of us and he had his Nissen hut at the end of Polbain, our net shed. And then we worked out of Old Dornie. We had a coble with an old Kelvin in, and also a Cathedral hull with a Volvo in. We had three or four nets out. One on the southern end of Isle Ristol, one over towards Badenscaillie and the third on the south side of Tanera Beg. And maybe another occasionally for a couple of years but you usually found that one net was pulling very well. Four was a hell of a lot of work though for two people. Nets get dirty and have to be changed. Then you get net damage. If you knew your net was getting a bit old, a bit sweet, you'd not work it hard.

There was a pause while he wandered off into the adjacent room to search for a photograph, which he didn't find!

See, I ended up working here from 1977, I think, to 91. But the thing was I was a single man here, living in a caravan, no running water, no electricity, no phone, no pull with an old car. So I used to get away most winters; Australia, New Zealand, working in the West Indies.

'Where were you born?' I had to ask!

On the Wirral, that's where I was born. You see, why I stayed with the salmon was that it suited my lifestyle. I'd start work in about April, getting it ready. We'd fish out by June and July and then by August, if I remember right, we were out of the water by then. But then you still have the nets to sort and dry and everything, and store away for the winter. Then because I was in the caravan, I would work for myself. People would say can you paint this, do that can you whatever. Then I'd disappear for the winter, I'd be gone and I'd come back in there spring and then I'd be doing work for people until the salmon started. I had this itinerant lifestyle and that was me for ten years or more.

It sounded to me a very similar life to Michael Barrett at Fascadale:

Basically all the salmon was run here by the Muirs and in the old days they were all local people, all crofters, so you had these guys who did their croft work and then they worked the salmon and they all fitted in together. When I was working here Willie used to take a lot of guys who were going to university. Students. He could employ twelve people as it could be a very prolific fishing at times.

The fish, it seems, was taken to Garve by lorry in Ken's time, and again put on the train for Inverness. But it was getting late and I wanted to get a quick drive up to Reiff before dark, even if it was really windy by then, to where the bothy remains derelict. Finally I wondered whether it had been a good way of life. 'Yes,' he said immediately.

> Had the salmon netting been a full-time permanent job then I think after a couple of months you'd be fed up. But with fishing in such a limited period, you are all the time out to get more fish, and then suddenly when you were beginning to feel, oh God, do I have to change another net; you know it all tailed off and you just got on with the rest of your life.

'Was it profitable?' I asked. There was a pause:

> I bought this house on the proceeds, but the thing is this house was knackered and it took me three years to renovate it. But the thing is, for ten years I had petrol to put in the car and alcohol to consume and the rest was a piece of piss because it didn't cost me anything. So when others were paying rent, I was living in a caravan so I knew no one was going to give me a mortgage. So I scraped the money to buy it.

That is the nature of fishing through the seasons. As I thought about this coast, I realised that there were more bag-nets up towards Lochinver at both Clachtoll and Culkein Achnacarnin, the former's bothy today housing a local museum. Salmon fishing ceased in 1994. In reality, I guess this whole coast, from Kintyre to Cape Wrath, and along the northern coast as we shall discover, was adequately covered in salmon fishing stations. Strangely, the Outer Hebrides were never renowned for salmon fishing.

The way Sandy MacLachlan, and now Ken, had described their lives, and not forgetting Michael Barrett in his book about Fascadale, I suddenly realised it might well have been one I'd have been very happy in too.

33

NORTH COAST, EAST COAST, AND HOME

So that was that. The end of the road. At Reiff the road did just that, ended just by the bothy. We didn't have any time to carry on north and, anyway, we'd gone well past Applecross, the original terminus of this adventure. I'd been along the north coast back in 2011 and there were only bag-nets and sweep-nets, and they'll only be repetition.

I think I forgot to mention that it was Sandy MacLachlan at Cuil Bay who had mentioned that there were sweep-nets at Kinlocheil and on Loch Leven, at the mouth of the River Coe, below Glencoe. And I'm sure there were many others, including some on the north coast of Scotland.

I'd spent a few days along the north coast on a couple of occasions, even if it's a hell of a drag up here. The scenery is outstanding, as is the serenity associated with quiet places. The roads are all single track, which puts many off (and confuses many more), and the weather can be drastic, although on the last occasion in 2011 the weather had been fab while storms raged down south. And in Bettyhill, a walk over to the sweep-net fishery on the south-east corner of Torrisdale Bay at the mouth of the River Naver is a must.

I recall reading the regulations from Section 2 of the Salmon and Freshwater Fisheries (Protection) (Scotland) Act 1951 in relation to the sweep-net:

(a) fishing for or taking salmon by net and coble means the use of a sweep-net, paid out from a boat, whereby the salmon are surrounded by the net and drawn to the bank or shore, provided that –

(i) the net and any warps are not made or held stationary, nor allowed to remain stationary, nor to drift with the current, but are both paid out and hauled in as quickly as practicable and kept in unchecked motion under the effectual command and control of the fisherman, for the purpose of enclosing the salmon within the sweep of the net and drawing them to the bank or shore;

(ii) no stakes, dykes, other obstructive devices or other nets are used in association with the net;

(iii) the water is not disturbed by throwing of stones or other objects, or splashing or other activity in order to drive salmon into the area to be swept by the net; and

(iv) the net shall not come within 50 metres of any other such net already being paid out or hauled, until the last mentioned net has been fully hauled in to the bank or shore.

But here in Bettyhill we seemed so far away from such legislation because the fishermen tend to regulate themselves to protect the salmon. This station dates from at least 1746 – the date of the Battle of Culloden – and last fished in 1992. Today the bothy remains, along with the ice house with its turf roof. Ice came from the nearby lochs and the fish, once boiled in vinegar and packed, went as far as Vienna and Paris. Again they had a man keeping watch for the salmon and supposedly he would shout '*Iomair*' at them, which was 'row' in Gaelic. I seem to recall that, on one visit, there was still a coble knocking about the old buildings. The kids were young then, and they played in the sand as I wandered about and absorbed the energy of the place. Maybe we had a picnic upon the grass by the old station.

The last foreman was Jock Mackay of Achina and he worked there for forty-three years. According to the information board, he recollected that on 2 July 1962, they caught 979 salmon, which had to be carried up the shore by hand barrow. The same year they landed 334 in one single sweep. And they reckoned that in those days they left the same number of fish behind as they landed.

Further east is the fishery at Armadale, one of the last to be worked. This was regarded to be at least 200 years old, though bag-nets were introduced in the middle of the nineteenth century. The boats were launched and landed from a slipway 100-plus feet down from above, and were only accessible via a steep path. The fish was brought up by a blondin, a cable crane that operated by hauling up and lowering down a hanging crane pulley when it was run back and forth up and down the suspended cable. I remember looking down upon the beach below and thinking, wow, that really was an achievement to get fish back up the cliff. It seemed a bloody long way down! James Mackay is the last owner of the fishing rights.

Moving east again, there was another bag-net fishery belonging to Simon Paterson at Strathy. The fishery closed in 2007 but was worked from Port Ghrantaich, where the slipway was built in 1902. Today the old salmon fishing station is a guest house called Salmon Landings, which opened recently. Again the salmon was brought up from below by blondin. Ken Lowndes tells me that there was a large hole in the grass somewhere nearby that opened up on the rocks below. Ice would be collected and thrown into the hole over the winter to stay frozen until removed from below in the season. God forbid anyone falling down the hole. In 2011 I wasn't aware of this otherwise I would probably have gone in search!

That same year I recollect photographing three fellows working a bag-net onto net poles at Castlehill, just east of Thurso. I guess they must have still been fishing.

Before we leave these coasts and have a wee look at the east, it seems fitting to give James Mackay of Armadale the last word for he is the chairman of the Scottish Salmon Net Fishing Association and, as such, has been involved with Marine Scotland. He told me:

> The Scottish Government, along with Marine Scotland, dealt a severe blow to what was a thriving industry for over two centuries. As an ancient way of fishing, this should have been seen as a protected historical way of life, one having giving major employment in rural areas. This is the only way the iconic wild salmon can be caught, and purchased legally. Scottish wild salmon has for over ten years held the P.G.I. status in protected food names like Stornoway black pudding and the Arbroath smokie. We have enquiries almost daily for wild salmon during the summer months and the answer of refusal causes great disappointment.
>
> The Scottish netting season starts around 11 February annually. In 1999, members of the Salmon Net Fishing Association of Scotland agreed to postpone the start of their season by six weeks on a voluntary basis to help spring stocks and this remained in place for sixteen years. However, after being ignored by Marine Scotland, the Association membership decided to withdraw this agreement. We didn't say we were going to fish within our legal right; we just removed the voluntary pledge.

Nets drying on the grass across the road from the bothy at Portsoy.

After this came to light, Marine Scotland and Scottish ministers passed legislation in 2015 to ban killing of all wild salmon until 1 April annually. When rich riparian angling groups took the case to the European Union to take Marine Scotland to task regarding this, the EU was threatening a fine for this infraction, so the 'GET OUT OF JAIL' card was to do what was spelt out for them: to get rid of netting salmon in Scotland. The dirty way this was executed was with a three-year prohibition, although they said this was likely to be two years.

Every netting station who put catch returns to Marine Scotland could apply to participate in paid scientific work over these three years and all would receive a fair compensation package. In 2016 the promised scientific work never took place and in 2017 I was awarded a so-called contract and ended up being paid less than I put into the project (safety equipment etc.). The scientific project failed because of seals constantly in the nets and, with no proper deterrent, it was impossible to catch fish alive to tag, while the authorities refused to issue a licence to manage rogue seals.

The compensation package was a pure disgrace for the netsmen and was a way by Marine Scotland to stop netting as cheaply as possible. The bigger operations got paid on track record but the smaller companies were ripped off with small pittances of money. Each company dealt with their own compensation but it was clearly unfair in owning a heritable title.

2018 was a dry summer with little salmon getting into the rivers. We were supposed to start netting again in 2019 but because of the previously dry year, it was decided we were not going to be allowed to fish. The Scottish coasts were loaded with fish but they could not be monitored because no coastal netting was allowed.

Prohibition was set against the EU infraction, although now the goal posts had shifted to conservation issues. Scottish Ministers have now decided to pay some netting companies with proven accounts a one-off payment but all fisheries can retain their right to fish for salmon and can do so on Scottish ministers' say-so for up to ten years.

Thus the netting stations stay quiet, with all their assets of gear, which can amount to hundreds of thousands of pounds, staying dead in sheds, sitting around for no other job.

Many erroneously consider the east coast of Scotland the home of salmon because, for the angler, the main salmon rivers empty out into the North Sea. The big four salmon rivers in Scotland are all found on this side: the Spey, Dee, Tay and Tweed. But there are dozens more such as the Ugie at Peterhead and North Esk at Monrose, both of which had, until recently, healthy commercial fishing stations.

Two places of salmon interest can be visited on this coast and both lie on the southern shores on the Moray Firth. The Tug Net Ice house and fishing station on the eastern bank of the mouth of the River Spey is one such example.

Ice houses started appearing in the seventeenth century and the Tug Net dates from 1830. However, pre-dating this is the fishing station, which was built by the Duke of Gordon & Richmond in 1768 to serve the rich salmon fishing in the lower River Spey. Owned by the Gordon Castle Estates, there were said to be some 150 men working here in the 1790s. Fish were stored briefly before being sold on, but once the ice house was built, they could be stored for much longer. Ice was obtained mostly from the ponds especially dug at the back of the fishing station in winter and broken up by pickaxe before being carried by horse and cart to the ice house. Once there, it was layered with sawdust and straw between to increase insulation and prevent it forming solid blocks. The ice house was built two-thirds below ground level and ice would remain solid right through the summer months.

Ice would mean salmon could be stored up to ten days before being shipped by sea to Aberdeen or London. With the arrival of the railway at Spey Bay, trains carried it. There's a flagpole that is said to have had the flag raised when a load was ready to be taken to the railway station.

A gang of fishermen worked with seven men and one salmon coble. Three would be in the boat, one rowing and two paying out the net, while the four on shore ensured the sweep-net drifted downstream properly. Cobles would be repaired, or even built, on site.

It has often been reported that more than 1,000 salmon were caught in one day but documented evidence is much scarcer! Though, with the nine-teenth-century servant class in London striking for being fed on too much salmon, and later having it written into their contracts that they would only receive it three times a week, the conclusion has to be that there was plenty of it about!

Railways were the lifeblood of salmon (and herring, of course) distribu-tion in the second half of the nineteenth century and, for many, continued right up to the 1960s, the era that is known today for the Beeching Cuts. We learned how the fishermen around Ullapool took their salmon down to Garve on the branch line to Kyle of Lochalsh but on the Moray Firth there was a line from Elgin, across the River Spey and east as far as Portsoy.

Portsoy – *Port Saoithe* in Gaelic, which might refer to *saoithe* meaning 'saithe', suggesting a fishing harbour – is our second stop and is only some 15 miles east as the crow flies from the Spey. Here there's an ancient harbour that dates back to the seventeenth century, which is in fact the oldest on the Moray coast. The so-called 'new' harbour dates from 1825. There was a railway extension to the harbour from the first railway to reach Portsoy from Banff in 1859 but which later closed a couple of years after the railway from Elgin arrived in 1886.

The Salmon Bothy is a museum of local heritage and fishing history and is housed in the former ice house, which dates from 1834. It's in the part of town called Seatown and faces Links Bay. The building is tall – there's a community space upstairs – and the three chambers have vaulted ceilings. Although there's much history of the port that is not salmon-related (trade, smuggling and industrial), there's a superb model of a bag-net as was fished by the six-man crew of the salmon fishing operation in Portsoy. It was Sandy MacLachlan who had made me aware of this and thus it necessitated a visit. Alongside are the vertical poles and horizontal stretchers, and cork floats, used in the bag-net. At the side of the ice house is the bothy itself, which has six bunks, and which is now used as a family history research area. There's a salmon coble sitting outside, as well as the old net poles across the road, alongside the loch. Salmon fishing here ceased in 1995. Nevertheless, it's well worth a visit.

From the Moray coast, after a detour to photograph the Ugie salmon smokehouse alongside the River Ugie at Peterhead, which is the oldest in Scotland and dates from 1585, we headed south. As we were somewhat nearby on the journey south, we stopped off at Nigg Bay, just south of Aberdeen, where the first bag-nets were said to have been set. Nothing much to see these days. From there down to St Cyrus, a few miles north of Montrose, where salmon fishing was practised using nets that were staked to the sandy beach. These were all but bag-nets except by name, the other exception being that, because of the nature of the beach, they uncovered at low tide and thus could be relieved of fish at that time instead of having to motor out in a coble. They were variously called fly-nets and jumper-nets, depending on the way they were set up. Indeed, the ice house is another relic from the salmon fishing days that has recently been converted as a dwelling now available for rent. Described as a 'luxurious retreat that is steeped in history', it has three bedrooms and costs in the region of £950 a week, although I guess this alters at different times of the year.

Just south again and the netting in the river at Kinnaber, on the North Esk, was one of the very last to cease fishing in recent years.

Drift-nets were set out at sea, usually close to the river mouths, and these captured salmon before they managed to swim into the migratory rivers. Then net and coble fisheries worked in the rivers and thousands of salmon were landed each week along this coast. It's no wonder I begin to feel sorry for this, the king of fish.

But, as we conclude, we mustn't ignore the herring for the east coast was much more active in herring than the west. Wick, at the north-east tip, was the herring capital of Europe in 1865 with millions of fish being barrelled in salt and exported all over. Seven miles south of Wick by road are the Whaligoe Steps, an amazing man-made stairway of 365 steps that descend to what was a naturally formed harbour between two sea cliffs – once a landing place for fishing boats. Thomas Telford surveyed the place in 1786 when considering where to put the herring stations of the British Fisheries Society (both Ullapool and Wick were decided upon, along with Tobermory on Mull and Uig on Skye). He regarded it as a 'terrible place', which it was. The steps date originally from the mid-eighteenth century and were once used by fisherwomen to haul up the creels of herring landed at the harbour beneath. Crews of women, some in their early seventies, would gut the fish and would carry them up the steps in baskets to be taken on foot to be sold. Barrels made in the cooperage at the top of the cliffs were taken down for salted herring to be stored in, to be exported by schooner.

Stretching right along this coast, from Wick to Berwick, the folk of small communities relied upon herring for incomes, and more than a thousand boats set out under sail, chasing the fish as they, along with the catching season, swam south and brought the catch home. Drift-nets were used, huge in length, although ring-nets did work specific areas. And, while we are on the subject of places to visit, we made our final stop at the Scottish Fisheries Museum in Anstruther, Fife, which is a must. Although the emphasis here is on herring, there's much about salmon too. Two great fisheries in one house. And if that isn't enough, then you have the east and south coasts of England to journey upon. Me, I'm off home!

JOURNEY END

At the journey's end then it is hardly surprising that, although I remain totally supportive of fishermen who have made their money from commercial fishing throughout their lives, I do feel a huge empathy towards the salmon, probably more so than the herring, although this sense of collective shoaling is exquisite. But what a magnificent creature the salmon is, both in appearance and achievement. Not only through his migration to the north and back, but the sheer remarkable fact that they say a salmon can smell the river in which it was born from 14 miles way. And we still know little about their migration habits. So you could say it's even more vital to ensure that the Atlantic salmon doesn't disappear completely from our seas. What an incredible journey this has been, to meet some wonderful people and discover some beautiful stories.

While there have been arguments between commercial and rod fishers for generations, it seems that the plug has finally been pulled on the former as the battle has been lost. Surely this is just part of the dire determination by some, especially at the higher, wealthy end of society, and supported by the current government, to see the balance of the liberalism of the 1960s and '70s rebalanced so that the overwhelming status of the 'one per cent' is returned to its 'rightful' place, leaving the yoiks left with the remnants of desperation – in other words, echoes of Margaret Thatcher's 'return to Victorian principles' ideology. While the criminalisation of trespass is under review, I read recently that I, and 99 per cent of you, are excluded from some 92 per cent of land and 97 per cent of waterways by the laws of trespass. Riverbanks constitute a good proportion of this.

At the same time, seemingly, I can't get farmed salmon away from my mind. Forget the loss of the traditional fishing methods for a moment. Consuming the wild stuff is limited to those landowners while we, the yoiks, are fed the insipid farmed stuff. Christ, there's irony there. But, by now, I would have hoped the reader will have made his or her own mind up.

We can't overlook the fact that there are many of us who do hark back, accepting that some of the ways of the past age do have their good points. One parting comment sticks in my mind: Willie Armer in Glasson Dock remarked thus:

> It was a way of life, and when you think, oh, maybe it's evolution or whatever you call it, but sadly I think we haven't moved on forward for the best. Like my mum said, we were poor but we were happy. And, er, you look now and what are people working for? It's mainly for materialistic things. You know, the outgoings for television licence, mobile phones, Sky TV, a better car, kitchen improvements, garden things, stuff like this, and so, consequently, to get by on stuff like that. As my dad used to say, 'You don't know how well off you are.' 'Well no, you were better off in a lot of ways,' I'd reply. 'What do you mean?' 'Well, when you look now, you didn't have cars, car tax to pay, cars to buy, car repairs, TV licences to buy, not the same income tax as there is now;' know what I'm saying. A lot of people are poor, but there again a lot of people have made their decision and gone away from natural little things, making a living off a little plot, many would be self-sufficient and wouldn't go and buy anything because they wouldn't need it ...

Of course, that's not true for everyone, especially in towns and cities. The urbanisation of this country, and now the make-up in the rural communities, has also changed, with second homes, retirement and commuting, but for those having spent a lifetime fishing out on the river to be suddenly stopped for reasons that don't add up, there would certainly be reflection on the ways of the past. I felt the same sadness seeing how Michael Wilson's market garden in Flookburgh was to become twenty-eight homes. All over in my travels, I've seen the effect of the boom in house building; from Appledore to past Applecross, they've sprung up at an alarming rate. Meanwhile, we import vast amounts of food as agricultural land lies dormant or is built upon. Fish from China, vegetables from Africa. How daft is that! And, from this book's point of view, all because we are told that fishermen are destroying salmon stocks, masking the truth that in fact it's purely the age-old reason: money talks. As prosperity increases on a hyperbolic scale and the comparable fortunes of many decline, woe betide anyone who gets in the way.

ACKNOWLEDGMENTS

My heartfelt thanks go out to the following, some of whom have since died, and are marked *

In Devon: Felicity Sylvester, Stephen & Sheila Taylor, Stephen Perham
In Minehead: Michael Morgan, Paul Date
In Stolford: Brendan Sellick*, Adrian Sellick
In Watchet: John Nash*
In Weston-super-Mare: Richard Reynolds, Dave Davis
On the River Severn: Deryck Huby, Don Riddle, Simon & Ann Cooper, Alan Osment*, Raymond* & Ann* Bayliss, Martin Morgan, Steve Tazewell
In Carmarthen: Andrew Davies, Raymond Rees*
On the River Cleddau: Alun Lewis
At Sunderland Point: Tom Smith*, Alan Smith, Trevor & Margaret Owen, Phil Smith, Dave Nash*
On the River Ribble: Ian Lythgoe, Paul Sumner, David Morrison, Russell Wignall, Tommy Threlfall, Andrew Porter
At Glasson Dock: Val Simpkin, Bernard Black, Willie Armer
At Flookburgh: Michael Wilson, Tessa Bunney
At Annan: John Warwick, Barry Turner, Tony Turner, Allan Watson
At Tarbert: Angus Martin, Willie Dickson, Archie Campbell
On Mull: David McLachlan, Gerard McLachlan, Malcolm Burge, Bryan Gibson, Jim Corbett, Sheena Walker, Mary Corbett, Jane Griffiths, Christine Leach, Douglas Canning, Sandy Brunton, Alastair Mackie, Neil Cameron
At Cuil Bay: Sandy MacLachlan
In and around Ullapool and beyond: Peter Muir, Ken Lowndes, James MacKay, Mark Stockl, Ally Macleod, Donald Macleod
In general, as always: Mike Craine*; Mark Horton for his foreword and Amy at The History Press for her continued support. Again, the family at home and Ana and Otis for joining me along the way.

All drawings in the book are by the author.

Drawing of a ring-net boat.

RINGER GLAD TIDINGS Ⅲ
HAULING IN A RING-NET   (DRAWING FROM PHOTOGRA

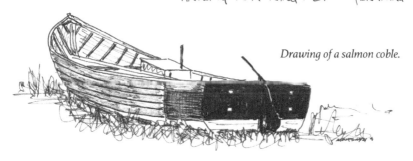

Drawing of a salmon coble.

SALMON COBLE   -2001

Drawing of the Castle Sod yair at Kirkcudbright.

WORKING THE YAIR

HAAF-NETTING

Drawing of the use of a haaf-net.

BIBLIOGRAPHY

Anonymous, *Observations Regarding the Salmon Fishery of Scotland* (Edinburgh: Bell & Bradfute, 1824).

Barrett, Michael, *The Leaper: Adventures in a Commercial Salmon Fishing Boat* (Indiana: Xlibris, 2016).

Bathgate, Thomas, 'Ancient Fish-Traps or Yairs in Scotland', *Proceedings of the Society of Antiquaries of Scotland*, 83, 7th series, 1948–49, pp.98–102.

Bradshaw, Jane, *Traditional Salmon Fishing in the Severn Estuary*, unpubl. dissertation, Univ. of Bristol, 1996.

Davis, F.M., *An Account of the Fishing Gear of England and Wales* (London: His Majesty's Stationery Office, 1937).

Dixon, John H., *Gairloch in North-west Ross-shire: Its Records, Traditions, Inhabitants, and Natural History, with a Guide to Gairloch and Loch Maree, and a Map and Illustrations* (Edinburgh: British Library, 2004).

Foster, Harry, *Don 'E Want Ony Shrimps* (Birkdale: The Birkdale and Ainsdale Historical Research Society, 1998).

Grimble, Augustus, *The Salmon Rivers of Scotland* (London: Kegan Paul, Trench, Trubner & Co., 1902).

Hawkins, James I., *The Heritage of the Solway Firth* (Annan, Friends of Annandale and Eskdale Museums, 2006).

Hayes, Nick, *The Book of Trespass: Crossing the Lines that Divide Us* (London: Bloomsbury Circus, 2020).

Hughes, Elizabeth, *A Glimpse of Old Handbridge* (Chester: 4 Corners Publishing, 2003).

Jenkins, J., Geraint, *Commercial Salmon Fishing in Wales* (Welsh Folk Museum, 1971).

Kissling, Werner, 'Tidal Nets of the Solway', *Scottish Studies*, Vol. 2, part 2, 1958).

Large, Nick, *The Glorious Uncertainty: Salmon Fishing in the River Severn* (Thornbury: Thornbury and District Museum, 2020).

Mackenzie, Murdo, *View of the Salmon Fishery of Scotland* (Edinburgh: Hansebooks, 1860).

Mackenzie, R., *Tales and Legends of Lochbroom* (Ullapool: Ullapool Bicentenary Committee, 1988).

Martin, Angus, *Fishing and Whaling* (Edinburgh: NMSE, 1995).

Mills, Derek, *Salmon and Trout: A Resource, its Ecology, Conservation and Management*, (Edinburgh: Oliver & Boyd, 1971).

Mitchell, Dugald, *Tarbert: In Picture and Story* (Falkirk: Birlinn Origin, 1908).

Smylie, Mike, *Herring: A History of the Silver Darlings* (Stroud: The History Press, 2004).

    Working the Welsh Coast (Stroud: The History Press, 2005).

    Fishing around Morecambe Bay (Stroud: The History Press, 2010).

Robertson, Iain, A., *The Salmon Fishers: A History of the Scottish Coastal Salmon Fisheries* (Ellesmere: Medlar Press, 2013).

Russel, Alex, *The Salmon* (Edinburgh: Edmonston and Douglas, 1864).

Walker, Jim, *By Net and Coble: Salmon Fishing on the Tweed* (Berwick-upon-Tweed: Blackwall Press, 2006).

Warwick, J. & Turner, B., *Annan Haaf-Nets*, annan.org.uk, 2018.

Wright, Sidney, *Curious Methods of Fishing in the World: Marvels of the World's Fisheries* (London: Logos Press, 1986).